Lecture Notes in Computer Science 5786

Commenced Publication in 1973
Founding and Former Series Editors:
Gerhard Goos, Juris Hartmanis, and Jan van Leeuwen

T0223476

Kurt Rothermel   Dieter Fritsch
Wolfgang Blochinger   Frank Dürr (Eds.)

Quality of Context

First International Workshop, QuaCon 2009
Stuttgart, Germany, June 25-26, 2009
Revised Papers

 Springer

Volume Editors

Kurt Rothermel
Wolfgang Blochinger
Frank Dürr
Institute of Parallel and Distributed Systems (IPVS), University of Stuttgart
Universitätsstr. 38, 70569 Stuttgart, Germany
E-mail:{kurt.rothermel, wolfgang blochinger, frank.duerr}@ipvs.uni-stuttgart.de

Dieter Fritsch
Institute for Photogrammetry, University of Stuttgart
Geschwister-Scholl-Str. 24D, 70174 Stuttgart, Germany
E-mail: dieter.fritsch@ifp.uni-stuttgart.de

Library of Congress Control Number: 2009934527

CR Subject Classification (1998): H.3, H.4, C.2, H.5, H.2, F.2

LNCS Sublibrary: SL 5 – Computer Communication Networks
and Telecommunications

ISSN   0302-9743
ISBN-10   3-642-04558-8 Springer Berlin Heidelberg New York
ISBN-13   978-3-642-04558-5 Springer Berlin Heidelberg New York

springer.com

© Springer-Verlag Berlin Heidelberg 2009
Printed in Germany

Typesetting: Camera-ready by author, data conversion by Scientific Publishing Services, Chennai, India
Printed on acid-free paper   SPIN: 12753727   06/3180   5 4 3 2 1 0

Preface

Advances in sensor technology, wireless communication, and mobile devices lead to the proliferation of sensors in our physical environment. At the same time detailed digital models of buildings, towns, or even the globe become widely available. Integrating the huge amount of sensor data into spatial models results in highly dynamic models of the real world, often called context models.

A wide range of applications can substantially benefit from context models. However, context data are inherently associated with uncertainty. In general, *quality of context information* has to be taken into account by both context management and applications. For example, the accuracy, completeness, and trustworthiness of spatial context information such as street or building data are very important for navigation and guidance systems.

QuaCon 2009 was the first international scientific meeting that specifically focused on the different aspects of quality of context data. Research in context management and, in particular, context quality, requires an interdisciplinary approach. Therefore, the QuaCon workshop aimed to bring together researchers from various fields to discuss approaches to context quality and to make a consolidated contribution toward an integrated way of treating this topic. We received 19 high-quality paper submissions by researchers from Europe, USA, and Asia. The International Program Committee selected 11 papers for presentation at the workshop. Additionally, five invited contributions by internationally renowned experts in the field were included in the workshop program. The presentations at the workshop showed many facets of quality of context from different research fields including context data management, spatial models, context reasoning, privacy, and system frameworks. The lively discussions underlined the great interest in this topic and in particular led to a deeper understanding of the relations between the various aspects of quality of context.

The success of QuaCon 2009 was the result of a team effort. We are grateful to the members of the Program Committee and the external reviewers for their thorough and timely reviews as well as to the authors for their high-quality submissions and interesting talks. We would like to extend special thanks to our invited speakers for their excellent and inspiring keynotes. Finally, we wish to thank all persons involved in the organization of the QuaCon 2009 workshop who did really a great job.

July 2009

Kurt Rothermel
Dieter Fritsch
Program Chairs
QuaCon 2009

Organization

QuaCon 2009 was organized by the Collaborative Research Center 627 "Spatial World Models for Mobile Context-Aware Applications" at the University of Stuttgart, Germany.

Executive Committee

Program Chairs	Kurt Rothermel and Dieter Fritsch
Organizing Chairs	Frank Dürr and Wolfgang Blochinger
Local Arrangements Chair	Michael Matthiesen
Financial Chair	Michael Matthiesen
Proceedings Chair	Wolfgang Blochinger
Demonstrations Chair	Frank Dürr

Program Committee

Reynold Cheng	University of Hong Kong, SAR China
Alois Ferscha	University of Linz, Austria
Andrew U. Frank	Technical University of Vienna, Austria
Christian Freksa	University of Bremen, Germany
Johann-Christoph Freytag	Humboldt-Universität zu Berlin, Germany
Dieter Fritsch	Universität Stuttgart, Germany
Hans Gellersen	University of Lancaster, UK
Christopher Gold	University of Glamorgan, UK
Michael F. Goodchild	University of California, Santa Barbara, USA
Christian S. Jensen	Aalborg University, Denmark
Bernhard Mitschang	Universität Stuttgart, Germany
Max Mühlhäuser	Technical University Darmstadt, Germany
Paddy Nixon	University College Dublin, Ireland
Sunil K. Prabhakar	Purdue University, USA
Kishore Ramachandran	Georgia Institute of Technology, USA
Kurt Rothermel	Universität Stuttgart, Germany

External Referees

N. Ahmed	M. Kost	L. Sun
A. Avanes	R. Lange	H. Weinschrott
M. Bhatt	K. Richter	D. Wolter
J. Chen	S. Rizou	L. Yusuf
L. Geiger	J. Shin	Y. Zhang

Sponsoring Institutions

Gesellschaft für Informatik (GI) e.V.
Informatik-Forum Stuttgart (infos) e.V.

Invited Speakers

Max Mühlhäuser, Technical University Darmstadt, Germany
Michael F. Goodchild, University of California, Santa Barbara, USA
Johann-Christoph Freytag, Humboldt-Universität zu Berlin, Germany
Reynold Cheng, University of Hong Kong, SAR China
Christopher Gold, University of Glamorgan, UK

Table of Contents

Invited Papers

Contributed Papers

Interacting with Context

Max Mühlhäuser and Melanie Hartmann

Technische Universität Darmstadt, Telecooperation Lab,
Hochschulstr. 10,
64289 Darmstadt, Germany
{max,melanie}@tk.informatik.tu-darmstadt.de

Abstract. Context is dodgy - just as the human computer user: hard to predict, erroneous, and probabilistic in nature. Linking the two together i.e. creating context-aware user interfaces (UIs) remains a great challenge in computer science since ubiquitous computing calls for lean, situated, and focused UIs that can be operated on the move or intertwined with primary tasks grabbing the user's attention. The paper reviews major categories of context that matter at the seam of humans and computers, emphasizing quality issues. Approaches to the marriage of context-awareness and user modeling are highlighted, including our own approach. Both sides of the coin are inspected: the improvement of UIs by means of quality attributed context information and, to a lesser extent, the challenge to convey context quality to the user as part of the interaction.

Keywords: Context-Aware Computing, Intelligent User Interfaces.

1 Introduction

Fifteen years after its advent, the term context-aware computing – and hence, the concept of context as used therein – remains a subject of definition quarrels. In the present paper, we apply an understanding that mediates between common usages of the term in the research literature (what kinds of systems and applications are called context-aware?) and etymological considerations (what would humans consider a correct use of the term 'context' in the light of its linguistic roots?). These considerations lead to the following definition:

Context in the sense of context-aware computing is

- *auxiliary* information[1] (in the sense of 'input'), used to *improve* the behavior of computer based systems (system or application software, appliances), whereby
- 'improved behavior' relates to the *functionality* itself or to the *user interface*
- and either *physical* conditions in which the software or system is operated
- or *human* factors related to the actual user(s)[2] are taken into account,
- both being predominantly *probabilistic* and *time-dependent* in nature

---

[1] This definition is at odds with, e.g., [28]. It excludes systems like navigation aids that *require* location as input yet reflects the meaning of 'in the context of' as 'in conjunction with'.

[2] Physical conditions and human factors as context were first distinguished in [29].

K. Rothermel et al. (Eds.): QuaCon 2009, LNCS 5786, pp. 1–14, 2009.

The QuaCon workshop rightly emphasizes the probabilistic nature of context, calling for approaches to the formalization and systematic consideration of this nature – and hence for elaborate concepts that cope with context *quality*. As to the second line item in the above context definition, the present paper is restricted to the role of context at the *user interface*. The paper title 'interacting with context' is deliberately ambiguous: both 'context-aware interaction' and 'user interfaces for conveying context information' shall be addressed in the remainder. Given these introductory remarks, the topic of the present paper can be formulated and structured as follows:

Interacting with context concerns		
	i) *improvement* of UIs based on *context,* including	
		quality (of context) *considerations*
	ii) *interaction* regarding context and its quality	

The italicized words in the above statement are used to devise the core structure of the paper. Chapter 2 addresses *improvements* i.e. ways to optimize UIs based on context. Chapter 3 deals with *context* proper: it discusses important categories and characteristics of context. Chapter 4 reviews *quality* 'types' and aspects of context, followed by approaches to the *consideration* of this quality in chapter 5, extending UI improvement as described in chapter 2. Chapter 6 turns to the second interpretation of '*interacting* with context', with a particular emphasis on how to communicate about and how to present the quality of given context information at the UI.

The aforementioned chapters concern the state of the art in general. Chapter 7 takes a look at specific approaches for interacting with context developed by the authors and their colleagues at our Lab, followed by conclusions and outlook in chapter 8.

2   Possible User Interface Improvements

We have to refine the distinction between system functionality and UI made in the above definition; as mentioned before, this paper focuses on the latter. Recent developments lead to the inclusion of context-aware information retrieval and information filtering (IR/IF) [1] in these UI considerations. At first sight, IR/IF is just an application domain, suggesting that context-aware IR/IF should just be subsumed under 'functionality'. However, as we can observe in our daily Web and desktop activities, IR/IF has become an integral part of most of our computer usages, tightly related to the 'how' and 'what' of UI usage (determination of parameters etc.). Therefore we split 'context-aware UI improvement' into context-aware *information handling* (in the general meaning of IR/IF) and context-aware *system use*, respectively. (Readers more interested in the *functionality* aspect of context-aware IR/IF may refer to [1].)

A. Information handling: with respect to IR/IF, context-awareness improves the information conveyed to the user. Three possibilities can be distinguished:

1. **Direct presentation:** context information proper is conveyed to the user; an example is the ContextPhone [2], which displays additional context information about a user, e.g. her current location, phone use activity, nearby acquaintances, etc.

2. **Indirect presentation:** context *influences* the information conveyed to the user; for this category, the Remembrance Agent by Rhodes [3] may serve as an example: it takes context information like the user's location and nearby people into account when selecting and presenting her formerly taken notes that might be relevant in the given situation; another often cited example category are tourist guides.
3. **Tagging:** context *influences meta data that influence* the information conveyed; this introduces another level of indirection, but also a more sustainable effect since meta data are subject to long-term evolution; we use the term tagging since it is the recently most wide spread term, in particular for social networks; other terms like profiling, semantic annotation etc. are in use, with slightly different meanings; an example is the ExpertRank and related StructuredTagCollection presented in [4].

B. System use: in this respect, context can improve either the 'outbound' behavior of a UI, denoted as *adaptation,* or its 'inbound' behavior, denoted as *execution.*

1. **Adaptation** concerns essentially the look-and-feel of a UI; different variants exist, in particular *reduction* and *adjustment*; it is particularly relevant if a UI shall be used with different I/O devices. E.g., if a UI 'migrates' from desktops to mobile devices, the amount of interaction elements displayed must be *reduced* and context information can facilitate the required choices. If the mobile device features touch screen capabilities, it is usually helpful to *adjust* the size of buttons, keys and other clickable objects to the usage context (e.g., use of styluses, fingers, or gloves);
2. **Execution** concerns the facilitation of actions performed by a user. Possibilities range from cautious suggestions like those 'hidden' in a context-aware ordering of options to audacious auto-execution of commands. For example, CHINLE [5] suggests which printing settings to use depending, e.g., on the file type to be printed; LookOut [6] suggests schedules for incoming meeting requests depending on the user's past behavior. This issue will be furthered later in the paper since it provides particularly good insight into the role of context *quality.*

The preceding paragraph alludes to different levels of system initiative. A study presented at Ubicomp 2003 [7] distinguishes three levels in this respect: *reactive* behavior means that the user has to adjust the context-unaware system herself (called 'personalization' in the study). At a first level of *proactivity* (called passive context-awareness), the system makes suggestions but asks for permission before adapting to the current context. A second level (active context-awareness) provides for automatic adaptation to the current context. The study provides evidence that users experience a lack of control when being exposed to passive or even active context-awareness but still prefer such proactive behavior over personalization due to the effort implied in the latter. Extrapolating this finding towards further improved concepts for proactive UIs (as can be expected from ongoing research) and towards an increasing number of computer-based systems used on the move (with very limited attention), we can safely assume that **proactive user interfaces** hold considerable potential. In conjunction with QuaCon and context-awareness, *proactivity* refers to all kinds of autonomous i.e. non user initiated (but maybe user acknowledged) context based improvements. In order to limit the users' lack-of-control and frustration experiences (e.g. due to un-wanted adaptations or even auto-execution), it is obviously crucial to consider the *quality* of the context information – this issue is investigated further in chapter 5.

3 Categories of Context – The Time-Dependent Nature

Rationale for this chapter. Literature about context-aware computing often shirks a clear definition of context, referring either to an arbitrary list of sample context types or to a literature reference – where the term is often ill defined, too. In [8], even the respected Anind Dey blames known definitions for relying on mere synonyms for the word context – and only one paragraph later, he provides a definition of his own that relies on the (roughly synonymous, again ill-defined) word *situation*.

A broadly accepted definition, taxonomy, and even model of context are not just 'nice to have'. Large-scale, efficiently-built, open collaborative (and hence, ubiquitous computing) systems can only be made context-aware as an ensemble if independently built parts can easily collaborate with respect to context information. This requires a common understanding (definition, taxonomy / categorization) and a common modeling approach. The latter must be carefully designed in order not to categorically exclude a certain kind of context or an aspect of its nature that is potentially relevant. The corresponding platform (middleware, toolkit or else) must be populated with immediate support for a significantly large number of context categories and aspects, including built-in context-processing as a service to software developers, and provide sophisticated means for extension since the hope for an all-inclusive list of context types is fatuous. Early context toolkits reflected only few categories of context and provided little support for their time-dependent and probabilistic nature (cf. [9]). The fulfillment of this important quest for a generic context model and platform is beyond the scope of this paper, but a few considerations about context categories shall be made in order to point at issues that are important for the theme of this paper. As a remark, the 'Mundo Context Server' developed at the Telecooperation Lab [10] is considered a contribution towards the fulfillment of this quest.

Is there a 'first-class context'? Location has often been considered as 'first-class context' (e.g., along with time and identity in [9]). Such a classification does not withstand careful investigation, neither in the sense that every second-class context would 'posses' a location context nor in the sense of another deterministic distinction. Location seems to be considered first-class because it is widely used and researched (cf. project Nexus [11]) – nevertheless, not every context is linked to a location.

Time is indeed useful in conjunction with virtually every other context type due to the time-dependent nature of context. However, in conjunction with other context information, peculiar *facets* of time act as *attributes* – but not as context information in its own right. Context-aware application may reflect a user's domicile, age, attention, and location, all of which exhibit a different degree and nature of time-dependency - this rules out the option to link them all to the same kind of 'time context'. Even strongly time-related potential context information like 'urgency' or 'time-of-day' cannot be computed on the actual time alone but must be linked to other context information (in the above examples to a deadline plus a workload or to a location, respectively). We therefore advocate considering time as an attribute of context rather than a context category. In conclusion, there should not be any distinction between first class and second class context.

In the light of a wealth of context models and platforms (cf. [12]) and efforts towards elaborate context ontologies (e.g., [13]), yet another categorization of context types may not seem appropriate. Nevertheless, the following considerations may help

i) to get a better understanding of the theme of this paper; ii) to identify white spaces in known efforts, and iii) to get one step closer to a common understanding of context.

A. Top categories – computational perspective. one important approach to the classification of context is driven by the question of how the context information is acquired. To this end, we distinguish three major possibilities:

1. **Physically sensed:** since the advent of cheap and multipurpose sensors and of wireless sensor networks was a catalyst for context-aware computing, an important (top-level i.e. broad) category of context concerns everything that can be technically sensed and (automatically) input to computer systems. Temperature, lighting, acceleration, and of course location are among the often-cited examples
2. **Virtually sensed:** context information may come from software sources outside the boundaries of interconnected context-aware platform and applications. A simple example is a calendar entry that includes a location: this location information may be just as valuable as a physically sensed one – and it is equally time-dependent and probabilistic in nature (a location sensor may be faulty or imprecise, a calendar entry may be outdated or attendees may be late).
3. **Derived:** it is common knowledge that (physically or virtually) sensed context usually needs processing in order to be useful for applications. A simple approach concerns the application of basic (arithmetic, lexical, ...) formulae for *filtering* and *aggregation,* yielding context information of roughly the same type as the sensed context: a *filter* may only pass the first 'person-X-in-room-Y' event of a series of identical such events (for person X), series of time-stamped keystrokes may be *aggregated* into keystrokes-per-second, etc. In contrast, the term 'higher level' context is often used to denote 'new' types of context computed from sensed ones – or again from derived ones. These new context types are either more appropriate for an application domain or closer to the users' assumed conceptual world. Two overlapping approaches prevail for deriving higher-level from lower-level context: *inference based* and *model based* concepts. *Inference* relates to the application of a rule set and to more or less sophisticated AI methods. *Models* are conceptual designs, mapped onto memory as interwoven data structures; the occurrence and 'aging' of context information may drive the dynamics of the model. The two approaches overlap since models may be manipulated using an imperative programming style or a declarative, rule-based one and hence 'by inference'.

B. Top categories – HCI perspective. Given our focus on context at the UI, it is helpful to consider UIs as the seam between humans and computers (or rather, systems). Interpreting the term system in a wide sense – including, e.g., the office, organization, business process, and social network in which it is used – any context that is not directly bound to the user can be subsumed under 'system context'. The next paragraph will discuss why a top-level distinction between **system context** (in this wide sense) and **user context** is helpful in conjunction with UIs.

B1. User context: Three context categories can be distinguished within this category:

1. **Emotional user context:** this category paves the way to the inclusion of fast growing research areas with a lot of potential for context-aware computing and

UIs, in particular affective (or: emotional) computing [14] and NLP based sentiment analysis [15].

2. **Cognitive user context:** several fields of HCI and in particular e-Learning lead to sophisticated user (learner) models, way beyond the simple models common in context-aware computing. Based on observations about a user's interactions with applications, they infer, e.g., goals and abilities. Since the 1980es frustration about over-hyped intelligent tutoring systems (ITS) has abated, substantial progress was made. In particular, so called open learner models bridge the gap between sophisticated modeling and easy-to-grasp user access to (and control over) the model [16] – a key requirement due to the *probabilistic* nature of context discussed below.

3. **Factual user context:** part of the user context can be sensed or rather easily derived from the environment; such context has been regularly used in context-aware computing and even before, e.g., in the form of simple user profiles. The present categorization shows that this 'first step' can be well blended with the other two – given a conceptually and technically integrated approach as demanded here.

Further details about these three categories must be skipped for the sake of space. However, the reader should be able to realize that context-aware computing can greatly benefit from a systematic inclusion of not just the third but also the first two categories (and respective approaches) – in particular at the UI where these fields have a large impact, but as of yet without systematic link to context-awareness.

B2. System context. Together with B1.3, B2 and its sub-categories reflect context taxonomies as known from the literature (cf. [12] again as a first entry). An analysis of the broad variety of previously published taxonomies suggests devising this category in a non-inclusive manner; *environment* and *activity* might be selected as initial categories, conforming to the recent proposals by [17] and [18].

In summary, we recall that the top-level categorization with respect to the *computational perspective* helps to discern simple i.e. sensed types of context and more 'high-level' context types (derived context) which are more appropriate for users and applications. The distinction into system and user context from the *HCI perspective* provides an approach to the grouping of these 'high-level' context types from the application perspective. The most crucial quest made in the present subchapter regards the top level distinction of user and system context and the harmonization of emotional and cognitive user context based approaches with context-aware computing.

4 Quality of Context – The Probabilistic Nature

In contrast to information sources like databases or user input, context is inherently probabilistic: sensors may fail, predictions of user goals may be incorrect, etc. Context-aware applications need to take this uncertainty into account when processing context. In order to be able to exchange context information between different components, these components need a common understanding about how context quality can be measured, modeled, and processed – the very topic of the QuaCon workshop. Buchholz et al. [19] propose to capture context quality as a list of properties:

- **Probability of Correctness** captures the core issue of how likely a given context is (initially) correct.
- **Trustworthiness** states the degree of trust put in the correctness of context; this property is added by an entity different from the context source, based on (historical) evidence; e.g., a hub node in a temperature sensor network may add this property to data (annotated with a probability-of-correctness) coming from different sensors.
- **Precision** captures the mismatch between context data and reality, e.g. +/-0.5° for a temperature sensor.
- **Resolution** describes the granularity of the context information, e.g. whether the temperature sensor measures the temperature for a room or for a building.
- **Up-to-dateness** captures the age of context data.

This list is an excellent inroad into the problem but does not address it in full. We will return to this issue, but first want use the list for substantiating our claim that user context can be further harmonized with other context categories. As an exemplary argument, we want to point out how all the above-listed 'quality of context parameters' can be provided for cognitive user contexts, too – except for *precision* in case of non-numerical values (where this measure usually makes no sense): i) algorithms used to infer information from the user's behavior usually return a confidence value along with the inferred information – this value can be considered equivalent to the *probability of correctness*; ii) *trustworthiness* can be derived in a similar way as for other context sources, e.g., by measuring how often a context source returned data that proved to be useful; iii) *resolution* may reflect the population from which the information was derived (single user, users group, all potential users, etc.); iv) the *up-to-dateness* can be reflected as the time of the last user model update.

Returning to the sufficiency of the property list, two major aspects should be addressed in addition; i) the level of **'statistical sophistication'** has a dramatic impact on how well context quality is reflected: while intervals or simple statistical parameters like average values provide for low resource demands, elaborate and more difficult to handle concepts like *pdf*s reflect reality much better; ii) **'model specifics'** i.e. further details and properties of the concept for context data capturing may be needed for proper context (quality) processing: e.g., a pressure sensor itself may be unable to determine precision but may provide additional temperature information (at the time of reading), used by a different component for computing precision *if* specifics about the sensor are known.

Quest for hierarchical quality details. There are forces pushing towards simple unified quality models, but also forces towards elaborate models. Three forces towards *simplification/unification* shall be mentioned: i) *resource economy* i.e. lower storage, communication, and computing demands (cf. average / interval based versus pdf based quality measures); ii) *interoperability:* e.g., by abstracting all quality properties to a single probability value, one can shield details of different positioning techniques from location aware applications, enabling the latter to operate on any such technique; iii) *cognitive load:* when conveying context information to the user (cf. chapter 6), simple, intuitive, easy-to-grasp concepts are essential. The key counterforce towards more *elaborate* context models is the processing of context quality: obviously, a simple probability value is less valuable than the list of properties above, and a simple 'capturing timestamp' is less valuable than the same value associated with an 'aging pdf'. Higher-layer components may be heavily dependent on details: e.g., a component for human activity detection may apply a machine learning algorithm that can greatly benefit from detailed sensor quality data, even without interpreting them and even if properties differ from one sensor group to another.

Obviously, both simplified/unified and elaborate models are needed. We therefore advocate the use of *hierarchical* quality of context measures. On the most coarsegrained level, a simple probabilistic measure for the overall confidence in a context value should be given, so that simple or independently developed components can easily incorporate context quality and that it can be easily conveyed to the user. On the next level, common quality of context properties should be given for a commonly agreed subset (the list above may be a basis, augmented by a formal semantic model). A further level may provide sensor / model specific additional quality measures. On a different dimension, a higher level should cover simple statistical values, whereas statistical details (e.g. pdf) should be given on a lower level.

5   Consideration of Context Quality in UIs

The probabilistic nature leads to inherent uncertainty of context information; this has to be considered when using it for improving the UI. E.g., no 'high-impact' or irreversible action should be executed automatically if it relies on uncertain sensor data. In the literature, three (non exclusive) meta-concepts are found that cope with uncertainty at the UI:

- **'inform and mediate':** inform the user about uncertain context and let her confirm or correct the data
- **'make multiple suggestions':** derive a weighted list of suggestions from context, not just a single one, and present them to the user for selection
- **'adapt behavior':** consider the level of uncertainty for adjusting whether and how an action is executed and suggestions are made

Using the *'inform and mediate'* meta-concept, the system explicitly asks the user for context information that cannot be sensed with sufficient reliability, or asks her to confirm context information or a derived action. Voice UIs exhibit a similar, wellknown behavior: users have to confirm their own input if it was not unambiguously recognized. However, asking the user for explicit confirmation introduces high cognitive load, especially if frequently applied. Therefore it is often advantageous to let the system apply context dependent actions immediately but to convey uncertainty information to the user (see next Section) and to enable her to modify the context information and the context-aware action. E.g., Mankoff and Dey [20] developed a system for mediating context information that is not unambiguously recognized.

The approach to *'make multiple suggestions'* is especially useful for information retrieval (cf. chapter 2). With this approach, the system suggests several items to the user that are derived from her current context. E.g., the Remembrance Agent by Rhodes [3] suggests conveying several notes to the user that match best to her current context. They are ordered by relevance with respect to the current context. This approach can be used in combination with *'inform and mediate'* as the system can suggest multiple context information to the user, asking her to choose the relevant one(s).

The third meta-concept *'adapt behavior'* is commonly used in the HCI community in the sense of adapting whether and how a user is addressed in relation to the confidence put in a context-dependent action. The approach is again intertwined with the other two. E.g., Horvitz uses three levels in his LookOut system [6] for scheduling meetings: level one takes no action, level two asks the user whether a computed appointment should actually be scheduled (cf. *inform and mediate*), level three automatically sets up an appointment for the user. This distinction is similar to the passive and active context-awareness discussed earlier [7].

All three meta-concepts mentioned have in common that they *refine and modulate the dialogue for interacting with the user*. This key principle should hence be considered when using context information for UI improvement.

6 Conveying Context Quality at the User Interface

One of the most important issues in building UIs is the establishment of user trust in the system she uses ([21], [22]). The user trust is difficult to gain and easy to lose again if the system takes (even few) erroneous decisions from the user's point of view [23]. Since erroneous conclusions cannot be avoided when using probabilistic context information, it is important to make the user aware of the quality of the context that lead to a conclusion. Thereby, users may get more realistic expectations about the system behavior, reducing the adverse effects of erroneous conclusions. Furthermore, Antifakos et al. [24] show that conveying the context quality can also improve the user's performance during interaction with context-aware systems. We can distinguish three main ways to convey context quality at the UI:

- **Numerical:** a numeric value (number) is used to represent the certainty in a given action / suggestion (e.g. [3], [25])
- **Symbolic:** different icons represent different levels of uncertainty; e.g., Maes et al. [21] use caricatures that indicate the confidence in a given prediction.
- **Gradual graphical attributes:** use a graphical attribute like color or line thickness to convey the certainty in the context quality. For example, CHINLE [5] uses different shades of green to mark which interaction element the user will most probably use next. Light green thereby means that the system is not very confident in the prediction whereas dark green stands for high confidence.

While it is important to convey the context quality as argued above, a drawback lies in the user's increased cognitive load. Therefore it is important that the context quality can be easily perceived and that it is conveyed in an unobtrusive manor. To this end, we advocate the use of gradual graphical attributes like colors (see below). Humans are used to this kind of quality depiction from everyday experience, making it intuitive to grasp. From our experience we advocate the presentation of one quality property at a time as opposed to various quality dimensions for the sake of simplicity.

7 The AUGUR Proactive UI Approach

In this section, we point out how the concepts discussed in this paper are applied in a proactive UI called AUGUR developed at the Telecooperation Lab. As can be expected, AUGUR aims at improving user interaction based on context information. AUGUR is able to augment any form-based web application with proactive support for entering data and navigation, without a need to modify the application. This provides for a large field of application of AUGUR, but of course limits its capabilities. Figure 1 shows an example screenshot of AUGUR as applied to the web page of the Deutsche Bahn AG (German railways). It shows how AUGUR integrates suggestions derived from context in the UI.

In terms of chapter 3, both system context (B.2) and cognitive user context (B.1.3) are taken into account. System context is derived from a universal context server developed at our lab [10], the cognitive user context comes from a user model that incorporates a learning component that reflects previous user interactions [26]. AUGUR uses this context information to support system use by facilitating the execution of actions. We distinguish three main *support types* for that purpose [27]:

- **Navigation shortcuts:** based on the occurrence of triggering events, AUGUR provides shortcuts to Web pages that the user might want to navigate to. E.g., a user may want to browse the details of a contact if this person calls. Figure 3a shows the suggestion as part of the AUGUR icon, which is integrated into each web page as a means for the user to interact with the AUGUR system itself.

- **Guidance:** users are guided through an application by highlighting the interaction element they will most probably interact with next. This reduces cognitive load and is especially useful for novice users or when navigating in large menu structures.

- **Content support:** input to one or several interaction elements is suggested according to the aforementioned meta-concept 'make multiple suggestions'. For each suggestion, AUGUR displays the corresponding context source(s). E.g., the content suggestions in Figure 1 are derived from the user's calendar ("Calendar Entry") and her current location (Berlin). If content support is provided for several interaction elements (cf. the second suggestion in Figure 1), data is filled in all these elements. AUGUR highlights all affected interaction elements, making the user aware of this automatic action. This kind of content support was shown to dramatically reduce the required interaction costs.

The three support types are presented to the user in a proactive way. In order not to annoy the users with too many proactive suggestions, AUGUR supports different levels of proactivity, depending on both the level of confidence and on the user's preferences. Three levels of proactivity are distinguished: AUGUR may …

- **highlight** elements to attract the user's attention (e.g., navigation shortcuts may be highlighted with a glowing AUGUR icon, providing for non obtrusive awareness)

- **suggest** data to be entered (visualized as a drop-down menu of interaction elements, ordered by confidence) or Web pages to switch to (visualized as speech balloons at the AUGUR icon, see Figure 3a)

- **automate** the execution of actions (prefilling input fields, auto-selecting data from drop-down menus, or automatically following links or clicking on buttons).

Fig. 1. AUGUR-enabled Web page of Deutsche Bahn AG (screenshot)

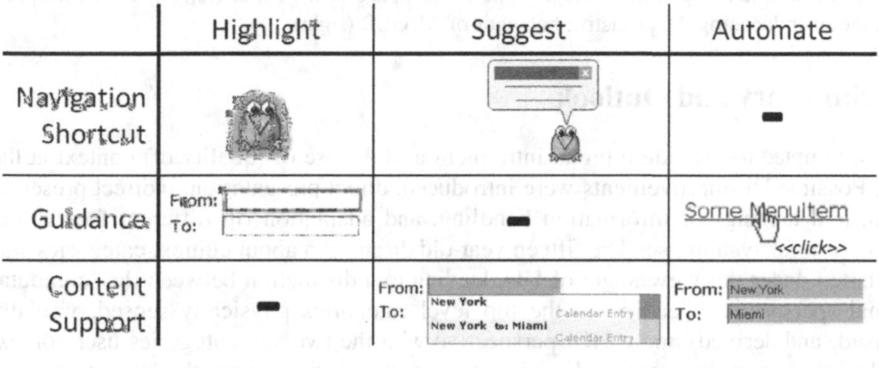

	Highlight	Suggest	Automate
Navigation Shortcut			—
Guidance	From: To:	—	Some Menuitem <<click>>
Content Support	—	From: To: New York New York to Miami	From: New York To: Miami

Fig. 2. Possible combinations of support types and levels of proactivity

Highlighting is used in support of automation: e.g., if content is automatically filled into forms, these fields are highlighted for user awareness and for indicating the option to change the content (cf. 'inform and mediate'). All combinations of support types and representations are summarized in Figure 2, indicating that not all these combinations are useful. Two combinations do not make sense (suggest for guidance and highlight for content support) and the third one, auto-navigation, is avoided in order to prevent user frustration due to serious 'lack of control'. In conjunction with guidance, auto-navigation is less dramatic but still risky; therefore, it is only provided if the user explicitly stated that AUGUR should perform this action if the confidence in it is sufficiently high.

In accordance with the meta-concept *'adapt behavior'*, the level of proactivity is adapted to the current confidence level for each support action of AUGUR. The confidence level is derived both from the quality of context information provided by the context server and from the confidence in the prediction, which in turn is computed based on a novel algorithm developed for AUGUR (details cf. [27]). The user can also adjust these settings by adjusting the minimal confidence for each level of proactivity required for a supporting action to be presented in that way (cf. Figure 3b).

In accordance with the arguments provided in chapter 6, AUGUR conveys quality of context information at the UI. The version currently in use applies very lightweight, easy-to-grasp means such as different shades of green if the support is presented using *highlight* as the level of proactivity and additional 'confidence level bars' if the *suggest* level is used.

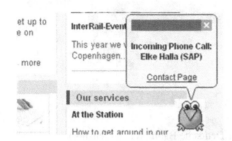 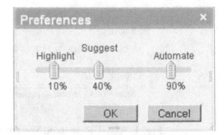

Fig. 3. a) Sample navigation shortcut to the contact page of the caller (left) **b)** User controlled thresholds influencing the proactive behavior of AUGUR (right)

8 Summary and Outlook

We attempted to provide a broad introduction to the use of (quality of) context at the UI. Possible UI improvements were introduced: direct presentation, indirect presentation, and tagging for information handling, and adaptation (in different facets) and execution for system use. The fifteen year old discussion about context categories was revisited under the view angle of UIs, leading to a distinction between the 'computational' perspective (leading to the top level categories physically sensed, virtually sensed, and derived) and a HCI perspective with the two sub-categories user context (split into cognitive, emotional, and factual) and system context, the latter forming a common ground for almost 'everything' known from other category models.

The discussion of *quality of context* provided further support for our quest to harmonize all three categories of user context with other context categories, and led to our suggestion of hierarchical details of quality. The issue of how to convey quality of context at the UI and a short review of our own system AUGUR – in relation to the issues treated before – concluded the main chapters.

The following outlook should have become evident from the article: fifteen years after the advent of the term context-aware computing, there is still a need for a sufficiently comprehensive context model and platform, which provides for sophisticated and holistic support, both of 'quality of context' and of UI related issues.

References

1. Brown, P.J., Jones, G.J.F.: Context-aware retrieval: exploring a new environment for information retrieval and information filtering. Personal and Ubiquitous Computing 5, 253–263 (2001)
2. Raento, M., Oulasvirta, A., Petit, R., Toivonen, H.: ContextPhone: A Prototyping Platform for Context-Aware Mobile Applications. IEEE Pervasive Computing 4, 51–59 (2005)

3. Rhodes, B.J.: The Wearable Remembrance Agent: A System for Augmented Memory. In: The Proceedings of the First International Symposium on Wearable Computers, ISWC 1997 (1997)
4. John, A., Seligmann, D.: Collaborative tagging and expertise in the enterprise. In: Proc. WWW (2006)
5. Chen, J., Weld, D.S.: Recovering from errors during programming by demonstration. In: Proceedings of the 13th international conference on Intelligent user interfaces Gran Canaria, Spain, pp. 159–168. ACM, New York (2008)
6. Horvitz, E.: Principles of mixed-initiative user interfaces. In: Proceedings of the SIGCHI conference on Human factors in computing systems: the CHI is the limit, Pittsburgh, Pennsylvania, United States, pp. 159–166. ACM, New York (1999)
7. Barkhuus, L., Dey, A.: Is context-aware computing taking control away from the user? In: Dey, A.K., Schmidt, A., McCarthy, J.F. (eds.) UbiComp 2003. LNCS, vol. 2864, pp. 149–156. Springer, Heidelberg (2003)
8. Dey, A.K.: Understanding and Using Context. Personal Ubiquitous Comput. 5, 4–7 (2001)
9. Salber, D., Dey, A.K., Abowd, G.D.: The context toolkit: aiding the development of context-enabled applications. In: CHI 1999: Proceedings of the SIGCHI conference on Human factors in computing systems, pp. 434–441. ACM, New York (1999)
10. Aitenbichler, E., Lyardet, F., Mühlhäuser, M.: Designing and Implementing Smart Spaces. In: Cepis Upgrade, vol. VIII, pp. 31–37 (2007)
11. Bauer, M., Becker, C., Rothermel, K.: Location models from the perspective of context-aware applications and mobile ad hoc networks. Personal and Ubiquitous Computing 6, 322–328 (2002)
12. Baldauf, M., Dustdar, S., Rosenberg, F.: A survey on context-aware systems. International Journal of Ad Hoc and Ubiquitous Computing 2, 263–277 (2007)
13. Chen, H., Finin, T., Joshi, A.: An ontology for context-aware pervasive computing environments. The Knowledge Engineering Review 18, 197–207 (2007)
14. Pantic, M., Sebe, N., Cohn, J.F., Huang, T.: Affective multimodal human-computer interaction. In: Proceedings of the 13th annual ACM international conference on Multimedia, pp. 669–676 (2005)
15. Inui, T., Okumura, M.: A Survey of Sentiment Analysis. Journal of Natural Language Processing 13, 201–241 (2006)
16. Susan, B., McEvoy, A.T.: An intelligent learning environment with an open learner model for the desktop PC and pocket PC. In: Artificial Intelligence in Education: Shaping the Future of Learning through Intelligent Technologies, p. 389 (2003)
17. Feng, Y., Teng, T., Tan, A.: Modelling situation awareness for Context-aware Decision Support. Expert Syst. Appl. 36, 455–463 (2009)
18. Paganelli, F., Giuli, D.: An ontology-based context model for home health monitoring and alerting in chronic patient care net-works. In: 21st International Conference on Advanced Information Networking and Applications Workshops AINAW 2007 (2007)
19. Buchholz, T., Küpper, A., Schiffers, M.: Quality of context: what it is and why we need it. In: Proceedings of the workshop of the HP OPenView University Association, HPOVUA. Geneva (2003)
20. Dey, A.K., Mankoff, J.: Designing mediation for context-aware applications. ACM Trans. Comput.-Hum. Interact. 12, 53–80 (2005)
21. Maes, P.: Agents that reduce work and information overload. Commun. ACM 37, 30–40 (1994)
22. Glass, A., McGuinness, D.L., Wolverton, M.: Toward establishing trust in adaptive agents. In: Proceedings of the 13th international conference on Intelligent user interfaces, Gran Canaria, Spain, pp. 227–236. ACM, New York (2008)

23. Leetiernan, S., Cutrell, E., Czerwinski, M., Hoffman, H.: Effective Notification Systems Depend on User Trust. In: Proceedings of Human-Computer Interaction–Interact 2001 (2001)
24. Antifakos, S., Schwaninger, A., Schiele, B.: Evaluating the Effects of Displaying Uncertainty in Context-Aware Applications. In: Davies, N., Mynatt, E.D., Siio, I. (eds.) UbiComp 2004. LNCS, vol. 3205, pp. 54–69. Springer, Heidelberg (2004)
25. Cheyer, A., Park, J., Giuli, R.: IRIS: Integrate. Relate. Infer. Share. In: Proc. of Semantic Desktop Workshop at the ISWC, Galway, Ireland (2005)
26. Hartmann, M., Schreiber, D.: Prediction Algorithms for User Actions. In: Proceedings of Lernen Wissen Adaption, ABIS 2007, pp. 349–354 (2007)
27. Hartmann, M., Schreiber, D., Mühlhäuser, M.: Providing Context-Aware Interaction Support. In: Proceedings of Engineering Interactive Computing Systems (EICS). ACM, New York (2009)
28. Schilit, B., Adams, N., Want, R.: Context-Aware Computing Applications. In: Proceedings of the Workshop on Mobile Computing Systems and Applications, pp. 85–90. IEEE Computer Society, Los Alamitos (1994)
29. Schmidt, A., Beigl, M., Gellersen, H.W.: There is more to context than location. Computers & Graphics 23, 893–901 (1999)

The Quality of Geospatial Context

Michael F. Goodchild

Center for Spatial Studies, and Department of Geography, University of California,
Santa Barbara, CA 93106-4060, USA
good@geog.ucsb.edu

Abstract. The location of an event or feature on the Earth's surface can be used
to discover information about the location's surroundings, and to gain insights
into the conditions and processes that may affect or even cause the presence of
the event or feature. Such reasoning lies at the heart of critical spatial thinking,
and is increasingly implemented in tools such as geographic information systems and online Web mashups. But the quality of contextual information relies
on accurate positions and descriptions. Over the past two decades substantial
progress has been made on the theory and methods of geospatial uncertainty,
but hard problems remain in several areas, including uncertainty visualization
and propagation. Web 2.0 mechanisms are fostering the rapid growth of user-
generated geospatial content, but raising issues of associated quality.

Keywords: geospatial data, context, uncertainty, error, Web 2.0.

1 Introduction

Over the past several decades there has been rapid and accelerating progress in the
availability, acquisition, and use of geospatial data, that is, data that associate places
on or near the Earth's surface $\mathbf{x}$, the attributes observed at those places $\mathbf{z}(\mathbf{x})$, and in
some cases the time of observation t. Progress can be seen in the development
of GPS (the Global Positioning System), which for the first time allowed rapid,
accurate, and direct determination of location; remote sensing, providing massive
quantities of image data at spatial resolutions as fine as 50cm; geographic information systems (GIS) and spatial databases to represent, analyze, and reason from
geospatial data; and a host of Web applications for synthesizing, disseminating, and
sharing data.

The purpose of this paper is to examine issues of quality when context is constructed from geospatial data. The next section provides some background, including a brief review of research on geospatial data quality and a summary of its major
findings. Section 3 examines the broader concept of context, drawing from work in
spatial analysis, the social sciences, and GIS. Section 4 discusses the key issues of
data integration, with particular emphasis on spatial joins and mashups. Section 5
examines the growing contributions of user-generated content, and the quality issues that are emerging in this context. The paper ends with a brief concluding
section.

K. Rothermel et al. (Eds.): QuaCon 2009, LNCS 5786, pp. 15–24, 2009.

2 Background

Early developments in GIS, and the automation of map-making processes, allowed information from maps to be converted to precise digital records. But paper maps are analog representations, and map-making is as much an art as a science, and it follows that data derived from maps do not necessarily stand up to the rigor and precision of digital manipulation, especially for scientific purposes. As early as the mid 1980s it had become apparent that the quality of geospatial data and the impact of quality on applications were significant and largely unexplored issues. A workshop in 1988 brought together the small community of researchers working on the problem, and led to a first book [1]. Two international biennial conference series were established in the 1990s (the 6th International Symposium on Spatial Data Quality meets at Memorial University, Canada, July 6–8 2009; and the 9th International Symposium on Spatial Accuracy Assessment in Natural Resources and Environmental Sciences meets at the University of Leicester, UK, July 20–23, 2010).

It quickly became apparent that the problem was much more than one of measurement error. The attributes associated with locations by ecologists, pedologists, foresters, urban planners, and many other scholarly and practitioner communities are frequently vague, with definitions that fail to meet scientific standards of replicability (asked to make independent maps of selected properties of an area, two professionals will not in general produce identical maps). Statistical approaches to error analysis were supplemented by research into fuzzy and rough sets, the theory of evidence, and subjective probability.

Today the field of geospatial uncertainty can be seen as addressing four related problems:

- sources of uncertainty, and approaches to uncertainty management and minimization;
- modeling of uncertainty for various types of geospatial data, using statistical and other frameworks;
- visualization and communication of uncertainty; and
- propagation of uncertainty during processes of analysis and reasoning.

Notable surveys of the field include those by Devillers and Jeansoulin [2], Foody and Atkinson [3], Goodchild and Jeansoulin [4], Guptill and Morrison [5], Heuvelink [6], Lowell and Jaton [7], Mowrer and Congalton [8], Shi, Fisher, and Goodchild [9], Stein, Shi, and Bijker [10], and Zhang and Goodchild [11].

Several key findings from this work can be identified. First, uncertainty should be defined as the degree to which a spatial database leaves a given user uncertain about the actual nature of the real world. This uncertainty may result from inaccurate measurement, vagueness of definition, generalization or loss of detail in digital representation, lack of adequate documentation, and many other sources. Second, uncertainty is endemic in all geospatial data. Third, the importance of uncertainty is application-specific, and may be insignificant for some applications; but it will always be possible to find at least one application for which the uncertainty of a given item of information is significant.

Measurement of geospatial position is never perfect, and may introduce uncertainty into the topological properties that can be derived from positions. For example, a point lying near the boundary of an area may appear to be outside the area if either its location, or the location of the boundary, or both are sufficiently uncertain. Similarly it may be impossible to determine accurately whether a house is on one side of a street or the other, because of uncertainties in the positions of both. Thus an important principle of GIS practice is that it may be necessary to *allow topology to trump geometry*, in other words to allow coded topological properties to override geometric appearances.

While the problem of uncertainty in geospatial data is in many ways analogous to problems of uncertainty in other data types, one key property leads to numerous fundamental problems. This is the property known as *spatial dependence*. Many types of errors in geospatial data tend to be positively autocorrelated; that is, errors of position x or attribute $z(x)$ tend to be similar over short distances. For example, suppose elevation has been measured at a series of points, with a standard error of 5m, and these elevations have been compiled into a *digital elevation model* (DEM) with a horizontal spacing between data points of 30m. A common application for such data is the estimation of slope. Clearly such estimates would be highly suspect if based on elevations with standard errors of 5m, if errors were statistically independent. In reality, however, methods of DEM compilation tend to create errors that are highly correlated over short distances. Thus it is still possible to obtain accurate estimates of slope despite substantial elevation errors.

A similar argument can be made for many applications of geospatial data. Databases of streets are useful for navigation purposes even though absolute positions may be in error by tens of meters, since *relative* positions are much more accurate. The area of land parcels can be estimated to fractions of a sq m even though their absolute positions may be in error by meters. Spatial dependence is the basis for the fields of geostatistics [12] and spatial statistics [13], both of which address the analysis and mining of spatially autocorrelated data. Informally the principle is known as Tobler's First Law of Geography [14]: "nearby things are more similar than distant things".

Several implications of the widespread presence of spatial dependence are worthy of mention. First, data that share lineage are likely to have spatially dependent error structures, and consequently *relative errors of positions and attributes will almost always be less than absolute errors*. In statistical terms relative error is a *joint* property of pairs of locations, whereas absolute error is a *marginal* property of locations taken one at a time. Second, when data from independent sources are brought together, with no sharing of lineage, relative errors will be as large as absolute errors. We return to this point later in the discussion of spatial joins and mashups.

The third implication concerns visualization. A map is a very effective mechanism for displaying the properties z associated with locations x, particularly when those properties are static. Measures of quality associated with locations, such as the marginal standard error of elevation discussed in a previous example, can also be displayed in this way. But the key issue of spatial dependence is problematic, since a map offers no way of showing the joint properties of locations, and thus no way of communicating to the user the important difference between correlated and uncorrelated errors. One solution, explored at length by Ehlschlaeger [15] and others, is to animate the map. For example, correlated errors of elevation will appear as a simultaneous rising and falling of neighboring points, like a waving blanket.

Finally, spatial dependence has implications for the data models used to represent geospatial data. Goodchild [16] has shown that the traditional model used to represent area-class maps (maps that partition an area into irregular patches of uniform class) cannot be adapted by adding appropriate attributes representing uncertainty to its various tables of nodes, edges, and faces; instead, an entirely new raster-based model must be adopted. Similarly, Goodchild [17] has argued that the traditional coordinate-based structure of GIS must be replaced by a radically different measurement-based structure to capture uncertainty in the measurement of positions.

3 Defining Context

Context can be defined as information about the surroundings of events, features, and transactions, and in the geospatial context of this paper surroundings can be taken to mean a geographic area. A host of terms have similar meaning, and in some cases those meanings have been formalized. Some of those terms and formalizations will be reviewed in this section.

Place has the sense of an area of the Earth's surface that possesses some form of identity, and perhaps homogeneity with respect to certain characteristics. Some places are officially recognized and formalized, such as the populated places recognized by the Bureau of the Census or the named places recognized by the Board on Geographic Names. Formalization often means the identification of a boundary, and often its digital representation as a polygon of vertices and straight connecting segments. A *gazetteer* is a relation between places, their locations, and their types [18], and the largest digital gazetteers currently contain on the order of 10^7 officially recognized places. Hastings [19] has discussed issues of geometric (locations), nominal (names), and taxial (types) interoperability among digital gazetteers. Other places have identity to humans, but no official recognition. Montello [20] discussed the place "downtown Santa Barbara", the elicitation of its geographic limits from human subjects, and the alternative representations and visualizations that would result from its formalization.

Community and *neighborhood* convey more of a sense of belonging. A resident at some location **x** would have some concept of belonging to an area A(**x**) surrounding **x**, and one would expect the neighborhoods of nearby residents to overlap substantially. In the extreme, one might expect a city to be partitioned into bounded and non-overlapping neighborhoods, such that all residents in a neighborhood perceive their areas A as identical. Increasingly, however, access to the Internet is creating communities that lack such simple geographic expression.

The *action space* of an individual is defined as the geographic area habitually occupied by the individual, including place of residence, workplace, and locations of shopping and recreation. Action space is clearly related to concepts of community and neighborhood, though many people would not identify workplace as part of neighborhood.

The idea that neighborhoods can be modeled as partitions lies behind the approach that many researchers have taken to unravelling connections between individuals and neighborhoods. For example, Lopez [21] has studied the impact of neighborhood on obesity, arguing that a resident's context determines his or her level of physical activity. Because of the difficulty of determining A(**x**) for every individual, researchers often assume that context is provided by the properties of some convenient statistical reporting zone containing **x**, such as a county, census tract, or block. Similar approaches have been used in studies of the effects of air pollution on health. In such cases context is easily accessible, but with obvious consequences of misrepresentation. Statistical reporting zones are rarely designed to coincide with anyone's sense of neighborhood, and the notion that all residents of a zone perceive the same zone as their neighborhood has little if any empirical support.

Geographers have long been interested in the partitioning of geographic space using formal criteria. A partition into *formal* regions is defined by minimizing within-region variation, while a partition into *functional* regions is defined as maximizing within-region interaction and minimizing between-region interaction, where interaction might be defined by patterns of trade, commuting, or social networking. In both cases the number of regions, and hence the average size of regions, must be determined independently.

Cova and Goodchild [22] have addressed the digital representation of A(**x**) when it is unique to **x**. They define an *object-field* as a mapping of location **x** to area A(**x**), and identify several other applications. In general this approach would be applicable to any problem in which context is unique to location.

More generally one might express context in terms of a convolution function. The context of a location **x** might be modeled as the aggregate effect of the properties of the surroundings, weighted by a function of distance w to allow nearby surroundings to contribute more than distant surroundings. If the surroundings consist of a set $\mathbf{y}_i$, characterized by some relevant attribute z_i, then context $C(\mathbf{x})$ might be defined as:

$$C(\mathbf{x}) = \sum_i z_i w(\|\mathbf{x} - \mathbf{y}_i\|) \bigg/ \sum w(\|\mathbf{x} - \mathbf{y}_i\|) \tag{1}$$

This approach has obvious advantages over the quick-and-dirty methods discussed previously. Rather than equating context with some independently defined reporting zone, it allows context to be defined explicitly through the function w, based on the spatial variation of some property z. Of the possible distance functions, the negative exponential has the advantage of being supported by extensive theory, showing that it is the most likely option in the absence of other information [23]. Negative powers have the disadvantage of $w(0)$ being undefined. In both cases however the weighting function will have a parameter, representing neighborhood scale, that must be established independently.

4 Data Integration

In geospatial technologies, location provides the common key to integrate data. In practice, however, there are substantial difficulties in doing so [24]. Lack of interoperability can be caused by differences of format, coordinate systems, or geodetic datums, and also by lack of documentation and vagueness of definition. In this section we consider the difficulties introduced by these issues, and techniques that have been designed to deal with them.

Standard methods of quantitative analysis, including virtually all of the methods of statistics, assume that data are arrayed in the form of a table—in database terms, they exist as elements of the tuples of a relation—with the variables occupying the columns and the rows corresponding to the cases. In a geospatial context, to analyze any relationship, such as that between the health of an individual and the pollution levels of that individual's neighborhood, it is necessary to define the variables of interest on a common *support*, in other words a common set of geographic features. In practical terms this means transforming the relevant variables to occupy columns in the same table, where the rows define the geographic features that provide the support. In this example, a GIS operation would be required to transfer the neighborhood pollution levels from their associated areas or polygons to the point support provided by the individual point locations. The *point in polygon* operation takes a set of points and a set of non-overlapping polygons, and identifies the containing polygon of each point.

This kind of operation is one form of *spatial join*, a conceptual extension of the relational join based not on a common key but on geographic location [25]. Equivalent operations exist for identifying all forms of intersection or containment between points, lines, areas, and volumes. But one characteristic distinguishing spatial joins from conventional relational joins is their uncertain nature. Blakemore [26] was one of the first to point out that the outcomes of the point-in-polygon operation could be in, out, maybe in, and maybe out, and that the option of a point lying exactly on a boundary was not feasible given the inherent uncertainty of position in any spatial database. Another common problem faced by any researcher dealing with line or area support is that two independently acquired versions of the same line or boundary will never agree, again because of positional uncertainty. This problem is commonly addressed by removing the inevitable slivers between different versions, replacing them by a single line.

It is useful at this stage to distinguish between locations specified by coordinates, such as latitude/longitude, and locations specified by placenames, street addresses, ZIP or postal codes, or other indirect methods. Today these various methods are essentially interoperable, and a number of Web services exist to transform between them. Digital gazetteers allow placenames to be converted to coordinates, while address-matching or geocoding services allow similar operations for street addresses. Point-of-interest (POI) services provide coordinates in return for business names, institutions, and many other features not normally found in digital gazetteers. Unlike coordinates, addresses, placenames, and POIs are nominal, so uncertainty arises in different forms when these references are used to execute spatial joins. For example, references to the same street address may vary because of syntax or abbreviation (compare 909 West Campus Lane with 909 W Campus Ln), placenames may vary because of multiple naming (compare Saigon with Ho Chi Minh City), whereas coordinates are subject to the kinds of uncertainty always associated with measurement on continuous scales. Further uncertainty arises because of the geographic extent of some features. For example, a placename such as Eiffel Tower may pin down a location to a few hundred meters, whereas Manhattan resolves location to no better than 10km.

GIS is often presented as a technology for integrating different *layers* of geospatial data, often conceptualized as maps of different types of geographic features. For example, GIS might be used to integrate point data about individuals with area data about neighborhoods, or area data about soil types with area data about current land use. Such operations are generally termed *overlay* because they involve a virtual superimposition of layers, as one might superimpose transparent maps. Integration in such cases can be interpreted in two distinct ways, however. A *topological* overlay is in effect a spatial join, transforming the contents of two or more layers to a common support, so that the contents of all layers can be compared in a single table. On the other hand a *graphical* overlay simply superimposes the layers in a visualization, allowing the user to see the spatial relationships and make inferences based on intuition. In a graphical overlay no transformation of data occurs.

Consider, for example, the Advanced Emergency GIS (AEGIS) developed as a joint project of the Loma Linda University Medical Center and ESRI, the leading developer of commercial GIS software. The system integrates and presents data relevant to an emergency via a standard Web browser, giving emergency managers an interactive situation overview. Clickable icons on the display depict many different types of features—roads, topography, cameras on the freeway network, the real-time locations of ambulances and helicopters, and the real-time locations of incidents of various kinds. This is a case of graphical overlay, no spatial joins having been executed. Thus it is the user's eye that positions an ambulance on a freeway, or a helicopter near a hospital, rather than any GIS operation based on the locations of these features. The map permits an effective level of management of the emergency, but it does not support more advanced applications, such as routing of ambulances from fires to hospitals, without the execution of appropriate spatial joins. Graphical overlay is comparatively unaffected by uncertainty of location, since the eye is capable of ignoring small errors in position when making many kinds of inference. However more advanced applications would require that positional uncertainty be addressed explicitly.

In the past few years the term *mashup* has become current in Internet applications, particularly of geospatial data. It refers to the creation of information from two or more Web sources, and derives from the practice in the music industry of mixing old tracks to create new ones. AEGIS can be described as a mashup, since the service it provides results from the integration of many independent data sources to create a new, useful product, based on geographic location as the common element.

Most mashups, including AEGIS, are examples of graphical overlay and do not require any form of spatial join. For example, the Housingmaps service (www.housingmaps.com) shows maps of properties currently for sale, by combining sales listings on the Craigslist service (www.craigslist.org) with Google Maps (maps.google.com). In this case the street address of the sales listing is converted to a coordinate reference before being positioned on the map, but no spatial join occurs, and it is the user's eye that identifies the context of a listing.

5 User-Generated Content

In recent years a new source of geospatial data has emerged through the use of so-called Web 2.0 technologies, which have enabled ordinary citizens to create their own maps and geospatial data. Hundreds of examples now exist, including some such as OpenStreetMap (www.openstreetmap.org) that offer a new alternative to traditional systems for the production of geospatial data, by inviting citizens to use GPS and other tools to create their own pieces of map, which are then integrated into a consistent patchwork. Wikimapia (www.wikimapia.org) invites users to contribute names, locations, and descriptions of familiar features, augmenting and to some extent replacing the services of digital gazetteers. Flickr (www.flickr.com) and comparable sites invite users to contribute georeferenced photographs and associated descriptions, producing a rich visual composite of the geographic world.

It is common to distinguish these new sources of geospatial data from the traditional sources such as the US Geological Survey (USGS), terming the latter *authoritative* and the former *asserted*. The providers of user-generated content typically have no qualifications or training in mapping, cartography, or geography, and rely instead on a plethora of easy-to-use tools such as GPS, the mapping software underlying services such as Google Maps, and their own familiarity with certain parts of the world. The term *neo-geographer* is often used in this context.

A number of authors have addressed the issues of quality and trust associated with asserted geospatial data [27]. Some services invoke elaborate sets of rules for identifying obvious errors in data contributed by users, whereas Flickr has no means of blocking a user who wishes to register a photograph in an obviously impossible location. In general research tends to find that levels of uncertainty in asserted data are no greater than levels of uncertainty in authoritative data, but that the mechanisms for assuring quality are very different. Traditional sources typically measure and publish the levels of uncertainty associated with their products, and require them to fall within the limits established by standards. Data quality is an important component of the *metadata* normally associated with authoritative data sets.

On the other hand metadata are conspicuously absent from most asserted sources, which rely instead on the concepts of *collective intelligence* or *crowdsourcing* to assure quality. In essence, it is argued that if enough people have the opportunity to review and edit information, that information will converge on the truth, or at least on a consensus. This principle tends to break down in the context of asserted geospatial data, however, since it implies that quality will be a function of the number of people interested in a certain item of information, and will therefore be best for information about areas familiar to the largest number of people, and poorest for areas that are remote, unfamiliar, or comparatively unpopulated. Moreover the notion of consensus may be problematic when dealing with types of geospatial data that are inherently vague.

It is tempting to believe that asserted data are inherently less certain that authoritative data. Agencies such as the USGS have accumulated a substantial public trust over the years, and even comparative newcomers such as Google tend to be trusted more than rank amateurs. But the standards that authoritative sources are required to honor do not guarantee perfection, only a maximum level of acceptable inaccuracy. The consequences of using authoritative data in a mashup can be very disappointing, even though the discrepancies fall well within published standards. Haklay has shown (povesham.wordpress.com/2008/08/07/osm-quality-evaluation/) that the uncertainties in the OpenStreetMap coverage of England are comparable to those of authoritative sources. In general, it appears that there is little difference between the uncertainty levels of authoritative and asserted data—rather, that uncertainties of authoritative data are well documented and guaranteed, whereas uncertainties of asserted data are largely unknown.

6 Conclusion

Although it is not the whole story, geospatial data clearly have vital importance in defining context. Research on defining the surroundings of an individual, event, or feature has a long and productive history, and in many cases progress has been made in formalizing the relevant concepts. Moreover there has been substantial and still-accelerating progress over the past few decades in systems for acquiring geospatial data, for compiling, documenting, and disseminating them, and for using them to define formally defined context.

Uncertainty is endemic in all forms of geospatial data, however, since it is impossible to capture the full richness of the Earth's surface and near-surface in the contents of any finite database. Effective models of uncertainty have been developed to deal with the particular characteristics of geospatial data, including spatial dependence, although much of this work remains mathematically complex and largely inaccessible to a wider audience. These uncertainties propagate from data to the evaluation of context, leading to errors, vagueness, and imprecision in defining exactly what surrounds a given individual, event, or feature. Moreover our techniques for communicating this uncertainty to the user, through visual or other means, remain very limited. The notion that maps can be inaccurate is largely alien to generations of humans who have grown up with the understanding that the Earth's surface is well mapped, and that there is consequently no need for maps to depict uncertainty. These same notions apply equally to the new world of map-like displays and mashups on laptops, PDAs, and cellphones as to the old world of paper maps and transparent overlays: visualization of geospatial uncertainty remains a hard problem, for both technical and cognitive reasons.

References

1. Goodchild, M.F., Gopal, S.: Accuracy of Spatial Databases. Taylor and Francis, London (1989)
2. Devillers, R., Jeansoulin, R.: Fundamentals of Spatial Data Quality. ISTE, Newport Beach, CA (2006)
3. Foody, G.M., Atkinson, P.M.: Uncertainty in Remote Sensing and GIS. Wiley, Chichester (2002)
4. Goodchild, M.F., Jeansoulin, R. (eds.): Data Quality in Geographic Information: From Error to Uncertainty. Hermes, Paris (1998)
5. Guptill, S.C., Morrison, J.L. (eds.): Elements of Spatial Data Quality. Elsevier, Oxford (1995)
6. Heuvelink, G.B.M.: Error Propagation in Environmental Modelling with GIS. Taylor and Francis, Bristol (1998)
7. Lowell, K., Jaton, A. (eds.): Spatial Accuracy Assessment: Land Information Uncertainty in Natural Resources. Sleeping Bear Press, Chelsea (1999)
8. Mowrer, H.T., Congalton, R.G. (eds.): Quantifying Spatial Uncertainty in Natural Resources: Theory and Applications for GIS and Remote Sensing. Sleeping Bear Press, Chelsea (2000)
9. Shi, W., Fisher, P.F., Goodchild, M.F. (eds.): Spatial Data Quality. Taylor and Francis, London (2002)

10. Stein, A., Shi, W., Bijker, W. (eds.): Quality Aspects in Spatial Data Mining. CRC Press, Boca Raton (2009)
11. Zhang, J.-X., Goodchild, M.F.: Uncertainty in Geographical Information. Taylor and Francis, London (2002)
12. Isaaks, E.H., Srivastava, R.: Applied Geostatistics. Oxford University Press, New York (1989)
13. Cressie, N.A.C.: Statistics for Spatial Data. Wiley, New York (1993)
14. Sui, D.Z.: Tobler's First Law of Geography: A Big Idea for a Small World? Annals of the Association of American Geographers 94, 269–277 (2004)
15. Ehlschlaeger, C.R., Shortridge, A.M., Goodchild, M.F.: Visualizing Spatial Data Uncertainty Using Animation. Computers and Geosciences 23, 387–395 (1997)
16. Goodchild, M.F.: Models for Uncertainty in Area-Class Maps. In: Shi, W., Goodchild, M.F., Fisher, P.F. (eds.) Proceedings of the Second International Symposium on Spatial Data Quality, pp. 1–9. Hong Kong Polytechnic University, Hong Kong (2003)
17. Goodchild, M.F.: Measurement-Based GIS. In: Shi, W., Fisher, P.F., Goodchild, M.F. (eds.) Spatial Data Quality, pp. 5–17. Taylor and Francis, New York (2002)
18. Goodchild, M.F., Hill, L.L.: Introduction to Digital Gazetteer Research. International Journal of Geographical Information Science 22, 1039–1044 (2008)
19. Hastings, J.T.: Automated Conflation of Digital Gazetteer Data. International Journal of Geographical Information Science 22, 1109–1127 (2008)
20. Montello, D.R., Goodchild, M.F., Gottsegen, J., Fohl, P.: Where's Downtown? Behavioral Methods for Determining Referents of Vague Spatial Queries. Spatial Cognition and Computation 3, 185–204 (2003)
21. Lopez, R.P.: Neighborhood Risk Factors for Obesity. Obesity 15, 2111–2119 (2007)
22. Cova, T.J., Goodchild, M.F.: Extending Geographical Representation to Include Fields of Spatial Objects. International Journal of Geographical Information Science 16, 509–532 (2002)
23. Wilson, A.G.: Entropy in Urban and Regional Modelling. Pion, London (1970)
24. Goodchild, M.F., Egenhofer, M.J., Fegeas, R., Kottman, C.A. (eds.): Interoperating Geographic Information Systems. Kluwer, Boston (1999)
25. Longley, P.A., Goodchild, M.F., Maguire, D.J., Rhind, D.W.: Geographic Information Systems and Science. Wiley, New York (2005)
26. Blakemore, M.: Generalization and Error in Spatial Databases. Cartographica 21, 131–139 (1984)
27. Elwood, S.: Volunteered Geographic Information: Key Questions, Concepts and Methods to Guide Emerging Research and Practice. GeoJournal 72, 13–244 (2008)

Context Quality and Privacy - Friends or Rivals?

Johann-Christoph Freytag

Institut für Informatik, Humboldt-Universität zu Berlin
Unter den Linden 6, 10099 Berlin, Germany
http://www.dbis.informatik.hu-berlin.de

Abstract. As our world becomes more and more proliferated by sensors
and mobile devices – often connected by wireless networks – there is the
urging need to develop appropriate abstractions for application development and deployment. Those abstractions should shield applications
from the physical properties of the devices thereby allowing applications
to focus on information processing based on global conceptual views (of
the world) in form of context models.

For the benefit of the user many mobile devices and applications communicate with others to perform their tasks. Therefore, we see the need
to give users some kind of control when and to which extend to allow
applications to communicate with other (mobile) devices or applications.
In particular, a user should be allowed to determine how much (s)he is
willing to share personal (private)data with others when participating in
such context aware infrastructures. That is, the user should have control
over how much his/her personal data is accessed by or communicated to
other systems if privacy is a concern to him/her.

This paper will elaborate on the concern for privacy in location-aware
systems by providing various examples that should highlight the complexity of such concerns. We show that privacy needs a well founded
base for handling user requirements appropriately. Often, privacy is not
a static property, but it is context sensitive thus increasing the overall
complexity of managing privacy according to the user's expectations. Additionally, we argue that quality aspects in context model based systems
should include and embed privacy protection and control mechanism as
an integral part on all levels therefore increasing the usability of such
systems from a user's point of view.

1 Introduction

Over the last decade the advances in sensing and tracking technology has led
to a proliferation of our world and lives with sensors and mobile devices often
connected by wireless networks. Today, applications on those devices such as cell
phones or personal assistants go beyond telephoning, messaging, or maintaining
address books as those devices often include various sensors. Those sensors might
create spatial information (such as information on the user's location or speed, or
information on horizontal/vertical orientation), environmental information (such
as information on temperature, noise, or brightness), or personal information

K. Rothermel et al. (Eds.): QuaCon 2009, LNCS 5786, pp. 25–40, 2009.
© Springer-Verlag Berlin Heidelberg 2009

(such as information on user's body temperature, the blood pressure, or pulse frequency). All these information items describe (part of) the current context of the user. This context might also be determined by other, non sensor created information such as user profiles or preferences, temporal information (such as time of day, season, year), social or activity information (such as "in a meeting", "interviewing", or "at a party"), resource information (such as information on available wireless access, battery status), or explicitly or implicitly stated user goals of current activities ("have to be in Stuttgart by 9 am").

These new sources of information create broad opportunities for new applications and application classes. Most of these applications support people and organizations to better manage their tasks and business related issues. For example, based on the user's current location a navigation systems might help the user reach his/her destination in a timely manner.

This recent development of ubiquitous devices and applications that access, combine, and transform context information from different sources has lead to the class of **context-aware systems**. Baldauf et al. [5] claim that one of the first context-aware systems developed was the *active badge aware system* by Want et al. [34]. They also trace back the term of context-aware systems to Schilit and Theimer who describe context as "... location, identities of nearby people, objects, and changes to those objects" [28]. Dey at al. gives a more general definition; they define context as "... any information that can be used to characterize the situation of entities (i.e. whether a person, place, or object) that are considered relevant to the interaction between a user and an application, including the user and the application themselves." [2]. Using context information as an important source for configuring and driving the system behavior has lead to the class of *context-aware systems*. Baldauf et al. review and classify existing approaches to context-aware systems before developing a generic architecture to design and implement such systems efficiently [5].

If these context-aware systems are predominantly location oriented we call them *location-aware systems*. Since location based information reflects and describes properties of real-world scenarios and situations it is important to develop *context models* that provide a general basis to interpret sensor based information in a coherent, consistent and meaningful manner. The NEXUS project is one example project whose goal is to "... to provide an infrastructure to support spatial-aware applications" [14] by developing "... methods and approaches for designing and implementing global and detailed (location-based) context models for mobile context-aware applications. Context models should include stationary as well as mobile objects of the real world. In addition, these objects should be complemented by virtual objects and services." (translated into English [26]).

When people use location-aware systems to support them in their tasks they usually take those around with them. Thus, these systems reveal location information about the user since the location information created by a sensor is identical with the location information about the user of such system. If, for example, the location of a device (and therefore of the user) is transmitted to another system (let it be a mobile or stationary system) this information might be essential to perform a user-requested task such as helping two people to meet or to generate a list of nearby restaurant. However, such information might also be used to the disadvantage of that user, either at the time of transmission – for example to send unwanted advertisement – or at a later point in time - for example, to determine that the user violated the speed limit while driving a car.

It is the goal of this paper to investigate these kinds of leakage of personal information from various points of view. The next section, Section 2, investigates the terms private information and privacy in general before Section 3 reviews existing approaches and systems to protect the privacy of users of such systems. Section 4 presents a list of general privacy principles that - from our point of view - should guide the development of any location-aware system that uses personal data to give the user the freedom and the control over private data that (s)he shares with other systems. Section 5 briefly introduces the EU-funded project PRECIOSA (**PR**ivacy **E**nabled **C**apability **I**n Co-**O**perative Systems and **S**afety **A**pplications) as an example of a location-aware system to demonstrate how to protect private information by various additions to the initial PRECIOSA architecture. Sections 6 and 7 provide a short list of challenges within the PRECIOSA project and a brief summary of the paper.

2 Private Data and Privacy

For the purpose of future discussions we first introduce and discuss the term sensitive and private data and the term privacy. Those are important to understand the technical challenges and threads that today's IT technology poses to the individual's right to privacy in various areas and systems, in particular in location-aware systems. According to the American Federal Standard FED-STD-1037C [4] its glossary defines *sensitive information* as information which through "... loss, or misuse, or unauthorized access to or modification of which could adversely affect the national interest or the conduct of federal programs, or the privacy to which individuals are entitled to under 5 U.S.C. Section 552a (the Privacy Act), ..." [36]. For purpose of personal context-aware systems we generalize the definition of *sensitive information* to mean

> information which through loss, or misuse, or unauthorized access to, or modification of which could adversely affect the interests of groups, organizations (such as the government or businesses), or the privacy to which individuals are entitled to by national or international law.

Obviously, this definition is very general, especially with respect to "interests of groups". We therefore limit the definition of sensitive information in the context of privacy to mean any information about a living individual. According to [16] "sensitive data means any information about a living individual that includes personal data revealing racial or ethnic origin, criminal record information, political opinions, religious or philosophical beliefs, trade-union memberships, and the processing of data concerning health or sex life". We call this kind of sensitive information *private (or personal) information* from here on. This previous definition resembles the definition of personal data as stated in the Directive 95/46/EC of the European Parliament and of the Council of 24 October 1995 on the protection of individuals with regard to the processing of personal data and on the free movement of such data [32], Article 2, Subsection a:

> *Personal data*[1] shall mean any information relating to an identified or identifiable natural person; an identifiable person is one who can be identified, directly or indirectly, in particular by reference to an identification number or to one or more factors specific to his physical, physiological, mental, economic, cultural or social identity.

As this definition already reveals, personal information relates to many aspects of an individual's life. It includes, for example, personal information on religious or sexual preferences, political or other personal activities (volunteer work in sensible social areas), information on sexual behavior or activities (for example using contraceptives or not) as well as information on financial matters. Accordingly, privacy has several facets that are context dependent. In general the literature distinguishes between

- *Physical (or bodily) privacy* which H. Jeff defines as preventing "intrusions into one's physical space or solitude" [30]. That is, physical privacy focuses on the protection of one's physical and emotional being when it comes to medical treatment by doctors, nurses, or medical institutions;
- *Sexual privacy* relating to sexual preferences, sexual activities, or other matters of the individual's sexual life;
- *Organization privacy* which relates to "Governments agencies, corporations, and other organizations may desire to keep their activities or secrets from being revealed to other organizations or individuals" [37];
- *Information (or personal) privacy* which relates to protecting *personal data* from being revealed to other organizations or individuals without the individuals consent.

In the following the presentation and discussion of this paper focuses on information privacy exclusively since the advances of IT technology over the last ten years in particular has dramatically impacted information privacy. Those technology advances come from more powerful hardware, improved Internet connectivity and therefore improved access to various data collections, better algorithms for data integration (fusion), and new sources of information that is easily accessible via the Internet; all of those advances pose an increasing thread to information privacy.

In 1998 Latanya Sweeney and Pierangela Samarati combined publicly available anonymized health care records of Massachusetts with voter records of the same state to reveal health conditions of individuals (including the former governor of Massachusetts). This experiment dramatically showed the power of IT technology even a decade ago. It did not only cause Sweeney and Samarati to

---

[1] Unfortunately, we recognize that this and many other references use the terms *data* and *information* as synonyms.

develop the notion of k-anonymity for protecting information from unwanted de-anonymization [27], but has also motivated and spawned several diverse research activities which are partially reviewed in the next subsection as far as location-aware systems are involved.

The right for personal privacy is continuously been challenged and continuously endangered even before massive advances in IT technology provided the means for "automated" privacy breaches on a large scale. Already in his book *Privacy and Freedom* published in 1967, Alan Westin was one of the first to define the concept of privacy in the context of modern communication infrastructures. For him privacy is

> the claim (right) of individuals, groups, or institutions to determine for themselves when, how, and to what extent information about them is communicated to others [35].

However, the term privacy was coined much earlier in the American legal system by the article of Warren and Brandeis in 1890 when large scale printing forced a clear standing of the law regarding individual's rights [7]. This definition has been the basis for many national and international regulations and guidelines such as the United States Privacy Act of 1974 [36], the Fair Information Practices [11], and the OECD Guidelines on the Protection of Privacy and Transborder Flows of Personal Data [24].

3 Privacy in Current Location-Aware Systems

Since the publication of the paper on *Protecting Privacy when Disclosing Information: k-Anonymity and its Enforcement through Generalization and Suppression* by Sweeney and Samarati [27] [31] there has been a swirl of efforts to develop various privacy protecting measures and to include those into systems. They protect sensitive data stored as records in a data collection by removing identifying attributes and their values from every data record (in database terms meaning removing key attributes and key values) and by forming groups of records whose quasi-identifiers — i.e. attributes whose values together might almost identify an individual — are forced to contain the same values.

A comprehensive overview on all aspects of privacy in different areas goes beyond the scope of this paper. However, we would like to mention that new concepts and implementations for protecting privacy have rapidly emerged in the areas of communication by using Chaum Mixes [8] or Onion Routing [12], and by using refined definitions of k-anonymity in database systems [31] [21] [20].

For the purpose of this paper we focus on the development of concepts and systems that appear in the context of location-aware systems. Many research efforts have focused on spatial-temporal query processing in (extended) relational database systems and the handling of moving objects by appropriate storage methods, query extensions and query processing frameworks, respectively. In both areas there is a large body of literature, all of which is relevant for location based systems.

Early approaches of handling privacy in location-aware systems were based on the idea of k-anonymization. Based on the user or system preference for the value of k the location-aware system would return an area (mostly a square like area) as the user's location which (s)he shares with k-1 other users (i.e. contains the location of k-1 other users). The advantage of this approach is its easy implementation using existing spatial data structures or indexes like the quad tree. The idea of cloaking by grouping one's own location with those of others has been used by several authors, i.e. for example by Gruteser and Grunwald [13], Chow, Mokebel, and Liu [10] and others [17]; it raises the important question how to manage the trade-off between preciseness of location information and the necessary level of anonymization to protect the user's privacy. Obviously, the preciseness of the user's location information is dependent on the number of other users around him/her thus impacting the usefulness of the location information for many location-aware systems. For example, when using a recommender system for restaurants or interesting places to visit, the user might be willing to tolerate some "fuzziness" in the result to protect his/her precise location (and therefore his privacy) as it might not impact the quality of the recommendations. However, for handling accidents involving cars or users the precise location information is important for providing immediate help to injured individuals possibly deciding about their lives or their deaths.

However, while the previous examples protect the data, it is equally important to protect the query – or better the user sending a query – from privacy breaches as well. For example when using a recommender system, it is important to know the user's current location; therefore it is often essential to send exact location information to the server together with the request for computing and returning a correct result. Besides knowing who sent the request (even this information might be necessary to disguise) we might use the same approach as described before. Instead of sending the exact location the user might send to the service an "approximate" location by sending some areal information that contains the user's location.

Although this approach might potentially suffer a loss of accuracy regarding the result, Mokbel, Chow, and Aref show in their New Casper system, that the *right level* of cooperation between the client and the server generates exact (location based) results for the user despite areal (i.e. anonymized location) information sent to the server to protect his/her location privacy [22]. That is, fuzzy local information does not automatically imply a loss of quality on the result. Their approach is to implement a *privacy-aware query processor* that generates an *approximate* result set containing all valid answers for a user submitted query. Based on the exact location information which is only known on the client system, a post-processing step (implemented either on the user device or by some middleware running at some trust site) filters out those result items that were returned due to the imprecise local information and thus do not belong to the answer for the user. For example, if the user asks for the nearest gas station, the server returns a set of potential gas stations based on the areal information;

the post-processing step then selects the correct one among all returned results based on the exact user location.

The New Caspar system uses geometric properties to handle private queries over public data, public queries over private data, and private queries over private data thus generating same results as without privacy measures. At the same time, the New Caspar system takes into account privacy profiles specified by the user [22]. The authors developed their system further resulting in the TinyCaspar which serves as a platform for generating alarm events based on individuals' location information without revealing those [9].

Other alternative approaches to protect the user's location privacy are based on hiding the user's identity [23] [6], by adding dummy location (i.e. location values for non-existing users) [19] [39] [38], or by transforming location data into an alternative space by a one-way function and answering queries in the transformed space [18].

Based on these different basic approaches several more complete systems have emerged whose goal is to implement a more complete and a more comprehensive solution to the problem of privacy in location-aware systems. The PAM system by Hu, Xu, and Lee allows to track as set of moving objects efficiently and precisely at the same time respecting the privacy requirements for those moving objects (if those represent people) [15]. The system requires knowing the queries in advance to allow the computation of so called safe areas. As long as the moving objects do not leave safe areas, the system does not have to recompute the answer to any of the (pre-registered) queries thus reducing the execution effort without compromising the quality of the answer.

4 Privacy Principles

The discussion on privacy enabling solutions for location-aware systems poses the important and challenging question if the solutions described above are sufficient to satisfy the needs for privacy protection. We argue that there exist several more aspects to information privacy that must be taken into account for a comprehensive solution. More specifically, we argue that we need guidelines or principles that describe in an implementation independent manner different aspects of privacy to be supported and to be maintained by any system. That is, those guidelines should describe specific functional needs for system designers and system implementers when building systems that deal with personal data. We shall call systems that adhere to those principles, privacy-aware systems. We argue that location-aware systems handle personal data and should therefore adhere to those principles.

For the purpose of this paper we use the following two publications as sources for designing those guiding principles for privacy and location-aware systems:

1. In their paper on *Hippocratic Databases*, Agrawal, Kirnan, Srikant, and Xu describe a set of principles that should guide any privacy-aware database management system (DBMS)[3]. We take the principles formulated in their

paper as a basis to slightly change them for the context of privacy and location-aware systems.

2. In its Subsection 7.2 the *"ISO Technical Report TC 12859 Intelligent transport systems - System architecture - Privacy aspects in ITS standards and systems"* provides recommendations "under which data shall be collected and held in support or provision of ITS services" [1].

The following principles clearly articulate in an implementation independent manner what it means for location-aware systems to access and to manage private information under its control responsibly. Furthermore, the principles also express for users what they can (and should) expect from the location-aware systems without being technically adept. The principles are as follows:

1. **Purpose specification:** for all personal information that is communicated between or stored by participating components of location-aware systems the purposes for which the information has been collected shall be associated with that information. The purpose might be explicitly or implicitly specified by the user or might be derivable from the current operational context. Furthermore, we expect the location-aware system to answer questions of the user such as why specific information is being operated on.

2. **Consent:** the user must provide his/her consent for the usage of information for a specified purpose. This consent might be restricted to one specific purpose or to a set of purposes; the user should also have the right and ability to revoke his consent for the future.

3. **Limited Collection:** Information related to the user (i.e. information describing properties or aspects of the user) shall be limited for communication, storage, and collection to the minimum necessary for accomplishing the specified purposes.

4. **Limited Use:** Any location-aware system shall execute only those operations on the personal information that are consistent with the purposes for which the information was collected, stored, and communicated.

5. **Limited Disclosure:** The personal information related to the user and operated on by a location-aware system shall not be communicated outside the location-aware system for purposes other than those for which the user gave his/her consent.

6. **Limited Retention:** Personal Information related to the user shall be retained by the location-aware system only as long as necessary.

7. **Accuracy and Context Preservation:** Personal Information related to the user and stored by the location-aware system shall always be accurate, up-to-date, and never be decoupled from its context and purpose.

8. **Security:** personal Information related to the user shall be protected by appropriate security measures against unauthorized use.

9. **Openness:** A user shall be able to access all information that is stored in the location-aware system and is related to the user.

10. **Compliance:** A user shall – directly or indirectly – be able to verify the compliance of the location-aware system with the above principles.

We briefly discuss the principles and explain dependencies among them.

- The *Principle of Purpose* is tightly bound to the information provided by the user; therefore this additional describing information (metadata) should not be separated as many of the other principles rest on that metadata. Of course this purpose specification is closely linked to the *Principle of Consent*. However, the *Principle of Consent* raises the question of how to resolve possible conflicts between the purpose specification and/or governmental or legal requirements that might contradict the user's consent.
- The *Principle of Limited Collection* requires deciding carefully which data is really needed to perform what kind of service. For example, if a user driving a car performs a hotel reservation, it might be necessary to provide a credit card number. However, it does not seem to be necessary to provide the current location of the car when making a hotel reservation.
- The *Principle of Limited Use* means to respect the purpose specification and the consent given by the user. This should be taken as one of the correctness criteria that need to be verified for a location-aware system under design or in execution.
- The *Principle of Limited Retention* again complements the purpose specification and the consent by the user. If the service or operation has been fulfilled there does not seem to be any reason to keep this information any longer. However, as already indicated before there might exist legal or governmental regulations that require such information to be kept longer than for the fulfillment of the purpose. Therefore, additional technical measures are necessary that help to obey both (possibly conflicting) requirements.
- The *Principle of Accuracy* seems to be controversial at a first glance since accuracy of information is closely related to integrity and quality of the location-aware system overall. However, we argue that inaccurate information also impacts the user's privacy – or as stated in the APEC Privacy Framework [29] "Making decisions about individuals based on inaccurate, incomplete or out of date information may not be in the interests of individuals or organizations".
- The *Principle of Security* requires technical measure that protect the user's information against "loss or unauthorized access, destruction, use, modification, or disclosure of data." [1].
- The *Principles of Openness and Compliance* is a particular challenge as a location-aware system in general is a distributed system where participating components might change over time. Therefore, specific technical measures might be necessary to fully comply with this principle for location-aware system. In particular, to verify compliance might be technically challenging as well as a costly task.

5   Privacy Awareness Implemented in PRECIOSA

The principles presented in the previous section look like a "wish list"; it is not clear per se if and how those are reflected in any system design or system implementation. To demonstrate the feasibility the author of this paper is involved in a two year European project called PRECIOSA (**PR**ivacy **E**nabled **C**apability In Co-**O**perative **S**ystems and **S**afety **A**pplications) [25] whose goal it is to determine concepts and approaches in the area of Intelligent Transportation Systems (ITSs), a specialized location-aware system, that realize the above ten principles for protecting privacy in ITS in an efficient manner. The "proof of concept" will be a prototypical implementation.

In the following, we briefly describe the design of one special ITS system under consideration in PRECIOSA which uses the future abilities of cars to access the Internet on a continuous basis. To study privacy issues in ITS we focus on a car based hotel reservation system that allows a driver or person in a car to make a hotel reservation by contacting a hotel reservation system (to make a reservation) and a credit card payment system (to secure the reservation), see Figure 1. The system uses a communication environment that consists of a client system (in the car), a road side unit (RSU) whose routing capabilities guarantee continuous connectivity with the Internet, and two server nodes for the two tasks, respectively.

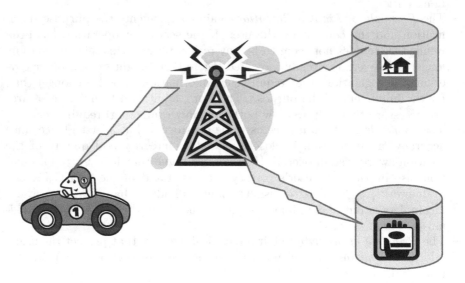

Fig. 1. The Car-Hotel-Payment Example

Obviously, there is an exchange of private data between the different nodes, all of which might be sources of privacy breaches. To ensure better privacy protection we envisage that each node of our (distributed) system relies on a basic, privacy-aware infrastructure. That is, the overall privacy-aware architecture for this application consists of privacy-aware ITS nodes (also called privacy-aware ITS components) that control the communication and the access to data. Each node executes its (part of the) application using a (standardized) privacy-aware ITS component that adheres to the privacy principles previously introduced. Figure 2 shows one instance of the privacy-aware ITS component with all its

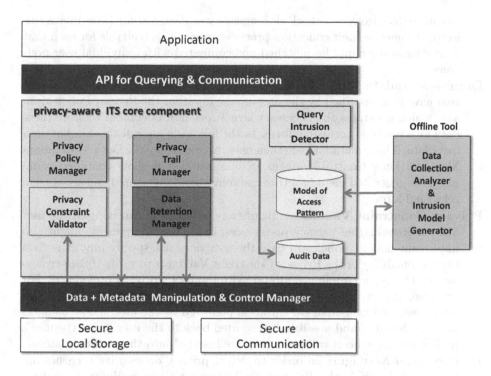

Fig. 2. A generic architecture for privacy-aware ITS

subcomponents. In the following we describe details of every subcomponent in Figure 2 and their relationships to the privacy principles as introduced in the previous chapter. To simplify the following discussion we assume

- that any application of privacy-aware ITS stores and accesses data via a query based interface of the privacy-aware ITS component. Similarly, all communication with other nodes is only possible via the privacy-aware ITS component;
- that the Secure Local Storage and Secure Communication allow to store data or to communicate data without privacy leakage, respectively
- that the Data + Metadata Manipulation & Control Manager is responsible for the compliance of the user's privacy preferences. That is, data is only stored, accessed, or sent according to the preferences expressed by the user.

In the following we briefly explain the functions and the goals of each of the components in Figure 2.

Privacy Policy Manager: Based on the above principles we must provide the data provider with means to express his/her privacy preferences by a privacy policy language. The expressiveness and the design of such policy language is currently an open issue. Examples for such a language could be the P3P language [33]. However, the P3P language was designed for (Web) data collectors to encode their collection practices in machine readable for such that P3P statements could be matched and compared with individual user preferences.

Data + Metadata Manipulation & Control Manager: Individual policies that have been specified by the user are the input for the Privacy Policy Manager which generates the necessary structures and its content to store these data in a machine accessible form. In the following we call this kind of data metadata. Those metadata are then made accessible by the Data + Metadata Manipulation & Control Manager. In [3] Agrawal et al. show a detailed example of the generated metadata that can conveniently be stored and accessed in relational form.

Privacy Constraint Validator: Before any application can be used the user must express his/her privacy preferences; individual components might also implement default policies in case the user does not specify any. It is then the responsibility of the Privacy Constraint Validator to verify if the privacy-aware ITS policies are in accordance with the user's privacy preferences. For example, the privacy-aware ITS might express a policy to store data for audit reasons longer than the duration preferred by the user. Those conflicts must be detected and possibly be reported back to the user. Once the user's preferences have been validated they are inserted into the metadata store.

Privacy Trail Manager: In order to detect privacy or security breaches on-line or off-line it is the Privacy Trail Manager's responsibility to collects and to analyze the sequence of events such as applications submitting a query or sending a message to another privacy-aware ITS component. Which information must be collected is up to a specific design.

Data Retention Manager ensures that data is deleted when necessary based on the preferences of the user. We realize that it might be necessary to make data inaccessible (i.e. logically deleting it) rather than deleting it if governmental or legal requirements exists. How to manage this trade-off and how to "seal data" such that it becomes inaccessible under regular terms must be worked out.

Data Collection Analyzer & Intrusion Model Generator is responsible for examining the operations either performed on the data store or on the communication link. Its goal is
- to control if any information is being collected (stored or transmitted), but not used in order to support the Principle of Limited Collection;
- to analyze if data is stored longer than necessary or specified by the user thus supporting the Principles of Limited Retention and Limited Use;
- to collect data on applications that access and communicate using the privacy-aware ITS component for detecting patterns that might suggest an intrusion of the privacy-aware ITS component;
- to collect data for auditing purposes; those are required by legal or governmental regulations. For generating intrusion models it might be necessary to cooperate with other components.

Intrusion Detector: Based on the patterns and suggestions generated by the Data Collection Analyzer & Intrusion Model Generator the Intrusion Detector is responsible for monitoring the privacy-aware ITS component – possibly in cooperation with other components – to identify those (possibly authorized) users that misuse the system or perform attacks to undermine the privacy protection measures by the privacy-aware ITS component or the privacy-aware ITS system as a whole.

The PRECIOSA project currently finishes its system design before focusing on the implementation of the system during the next twelve months.

6   Analysis and Future Challenges

The simplified, yet realistic example introduces an architecture that must be accompanied by additional features to specify and to implement privacy policies. Those policies are either set by default for each component in a privacy-aware ITS, or are set individually by users who express their personal preferences when using such system. In the following we briefly describe the three important challenges that need further discussion: the role of metadata, the timely enforcement of policies, and the concept of system privacy.

Data and Metadata: The principles for ITS and location-aware systems in general together with the above example clearly show that user data must be accompanied with additional data that go beyond determining possible domain values or structural properties when enforcing privacy. In the context of privacy-aware ITS those kind of metadata must be extended to guarantee the proper access and dissemination of data within such as system. Obviously, the source of this kind of metadata is the result of analyzing privacy preferences expressed by a privacy policy language and deriving those metadata from expressions of those languages.

Privacy Policy Enforcement: Enforcing privacy requirements in a privacy-aware ITS must be embedded in the overall life cycle designing and building privacy-aware systems. Therefore, privacy policies must be enforced at different phases of the system development process:

- During the design and implementation phase,
- During the deployment phase of the system,
- During the execution phase of ITS components.

Simple examples show that it is important and necessary to pay attention to privacy issues in all these three software development phases. However, it is beyond the scope of this paper to discuss more details.

System Privacy: Since privacy-aware ITSs are distributed systems consisting of several components and an underlying (communication) network it becomes especially important to understand how to build such a system from basic components. Similar to building correct distributed systems from correct components (independently of the definition of correctness), we must understand and ensure how to build privacy-aware ITS from individual components that already exhibit privacy properties that are known. That is, the composition of a privacy-aware ITS from different components must include a clear process that derives and guarantees a level of privacy for the privacy-aware ITS based on the privacy properties of the individual components. Only such a compositional approach will guarantee that the overall system enjoys a verifiable level of privacy.

7   Summary and Acknowledgment

In a world of ubiquitous computing, mobility, and increased compute power it becomes more and more important to build privacy-aware systems. In this paper we argue that privacy is an especially important aspect that has been discussed and worked on in various research efforts for location-aware systems.

To lay the foundation we reviewed the general notion of privacy before presenting ten general principles for protecting the privacy of individuals based on ongoing work of the technical and non-technical community. These principles should give guidelines how to design and implement privacy and location-aware systems. As an example, we use the EU-funded PRECIOSA project to demonstrate the challenges of implementing the ten principles of privacy effectively and efficiently. Although there seems to be a trade-off between privacy concerns and some quality properties in general, some research results show that such trade-off is not inherent.

I would like to thank all members of the PRECIOSA team for a creative and stimulating project environment. In particular, I am grateful to Frank Kargl (University of Ulm), Antonio Kung (Trialog, Paris), and my assistants Martin Kost and Lukas Dölle for many fruitful discussions, creative ideas, and for being reliable project partners.

I am also thankful to Bernhard Mitschang and Frank Kargl for a careful review of this paper. Despite their help I am responsible for all errors that still might be found in this paper. This work was supported by the EU project PRECIOSA.

References

1. ISO TC 204/SC/WG 1. Intelligent transport systems – system architecture – privacy aspects in its standards and systems. Technical report, ISO (2008)
2. Abowd, G.D., Dey, A.K., Brown, P.J., Davies, N., Smith, M., Steggles, P.: Towards a better understanding of context and context-awareness. In: Gellersen, H.-W. (ed.) HUC 1999. LNCS, vol. 1707, pp. 304–307. Springer, Heidelberg (1999)
3. Agrawal, R., Kiernan, J., Srikant, R., Xu, Y.: Hippocratic databases. In: 28th VLDB Conference, Hong Kong, China, pp. 143–154 (2002)
4. ATIS. Telecommunications: Glossary of telecommunications terms (2000), http://www.atis.org/glossary/default.aspx (accessed May 13, 2009)
5. Baldauf, M., Dustdar, S.: A survey on context-aware systems. International Journal of Ad Hoc and Ubiquitous Computing, 263–277 (2004)

6. Alastair, R.: Beresford and Frank Stajano. Location privacy in pervasive computing. IEEE Pervasive Computing 2(1), 46–55 (2003)
7. Brandeis, L., Warren, S.: The right to privacy (1890),
   http://dx.doi.org/10.2307/1321160 (accessed May 13, 2009)
8. Chaum, D.L.: Untraceable electronic mail, return addresses, and digital
   pseudonyms. Commun. ACM 24(2), 84–90 (1981)
9. Chow, C.-Y., Mokbel, M.F., He, T.: Tinycasper: a privacy-preserving aggregate
   location monitoring system in wireless sensor networks. In: SIGMOD 2008: Proceedings of the 2008 ACM SIGMOD international conference on Management of
   data, pp. 1307–1310. ACM, New York (2008)
10. Chow, C.-Y., Mokbel, M.F., Liu, X.: A peer-to-peer spatial cloaking algorithm for
    anonymous location-based service. In: GIS 2006: Proceedings of the 14th annual
    ACM international symposium on Advances in geographic information systems,
    pp. 171–178. ACM, New York (2006)
11. Federal Trade Commission. Fair information practice principles,
    http://www.ftc.gov/reports/privacy3/fairinfo.shtm (accessed May 12, 2009)
12. Goldschlag, D.M., Reed, M.G., Syverson, P.F.: Hiding routing information. In:
    Anderson, R.J. (ed.) IH 1996. LNCS, vol. 1174, pp. 137–150. Springer, Heidelberg
    (1996)
13. Gruteser, M., Grunwald, D.: Anonymous usage of location-based services through
    spatial and temporal cloaking. In: Proceedings of the ACM MobiSys, New York,
    NY, USA, pp. 31–42 (2003)
14. Hohl, F., Kubach, U., Leonhardi, A., Rothermel, K., Schwehm, M.: Next century
    challenges: Nexus—an open global infrastructure for spatial-aware applications. In:
    MobiCom 1999: Proceedings of the 5th annual ACM/IEEE international conference
    on Mobile computing and networking, pp. 249–255. ACM, New York (1999)
15. Hu, H., Xu, J., Lee, D.L.: Pam: An efficient and privacy-aware monitoring frame-
    work for continuously moving objects. IEEE Transactions on Knowledge and Data
    Engineering (to appear, 2009)
16. International Screening Solutions (ISS). Privacy terminology (1890),
    http://www.intlscreening.com/resources/terminology/
    privacy-terminology/ (accessed May 13, 2009)
17. Kalnis, P., Ghinita, G., Mouratidis, K., Papadias, D.: Preventing location-based
    identity inference in anonymous spatial queries. IEEE Transactions on Knowledge
    and Data Engineering 19(12), 1719–1733 (2007)
18. Khoshgozaran, A., Shahabi, C.: Blind evaluation of nearest neighbor queries using
    space transformation to preserve location privacy. In: Papadias, D., Zhang, D.,
    Kollios, G. (eds.) SSTD 2007. LNCS, vol. 4605, pp. 239–257. Springer, Heidelberg
    (2007)
19. Kido, H., Yanagisawa, Y., Satoh, T.: An anonymous communication technique
    using dummies for location-based services. International Conference on Pervasive
    Services, 88–97 (2005)
20. Li, N., Li, T., Venkatasubramanian, S.: t-closeness: Privacy beyond k-anonymity
    and l-diversity. In: ICDE, pp. 106–115. IEEE, Los Alamitos (2007)
21. Machanavajjhala, A., Gehrke, J., Kifer, D., Venkitasubramaniam, M.: l-diversity:
    Privacy beyond k-anonymity. In: ICDE, p. 24 (2006)
22. Mokbel, M.F., Chow, C.y., Aref, W.G.: The new casper: Query processing for
    location services without compromising privacy. In: VLDB, pp. 763–774 (2006)
23. Myles, G., Friday, A., Davies, N.: Preserving privacy in environments with location-
    based applications. IEEE Pervasive Computing 2(1), 56–64 (2003)

24. OECD. Oecd guidelines on the protection of privacy and transborder flows of personal data,
    http://www.oecd.org/document/18/0,3343,en_2649_34255_1815186_1_1_1_1, 00.html (accessed May 12, 2009)
25. Preciosa. Preciosa — **PR**ivacy **E**nabled **C**apability in co-**O**perative systems and **S**afety **A**pplications (2009), http://www.preciosa-project.org/ (accessed May 12, 2009)
26. Rothermel, K., Ertl, T., Fritsch, D., Kühn, P.J., Mitschang, B., Westkämper, E., Becker, C., Dudkowski, D., Gutscher, A., Hauser, C., Jendoubi, L., Nicklas, D., Volz, S., Wieland, M.: SFB 627 - Umgebungsmodelle für mobile kontextbezogene Systeme. Inform., Forsch. Entwickl. 21(1-2), 105–113 (2006)
27. Samarati, P., Sweeney, L.: Protecting privacy when disclosing information: k-anonymity and its enforcement through generalization and suppression. Technical report, Computer Science Laboratory, SRI International (1998)
28. Schilit, B., Theimer, M.: Disseminating active map information to mobile hosts. IEEE Network 8, 22–32 (1994)
29. APEC Secretariat. Apec privacy framework. Technical report, Asian Pacific Economic Cooperation (APEC) (2005), ISBN: 981-05-4471-5
30. Smith, H.J.: Managing privacy: information technology and corporate America. UNC Press (1994)
31. Sweeney, L.: Achieving k-anonymity privacy protection using generalization and suppression. International Journal of Uncertainty, Fuzziness and Knowledge-Based Systems 10(5), 571–588 (2002)
32. European Union. Directive 95/46/ec of the european parliament and of the council of 24 october 1995 on the protection of individuals with regard to the processing of personal data and on the free movement of such data (1995),
    http://www.cdt.org/privacy/eudirective/EU_Directive_html#HD_NM_28 (accessed May 13, 2009)
33. W3C. P3P — platform for privacy preferences project (2007),
    http://www.w3.org/P3P/ (accessed May 12, 2009)
34. Want, R., Falcao, V., Gibbons, J.: The active badge location system. ACM Transactions on Information Systems 10, 91–102 (1992)
35. Westin, A.: Privacy and Freedom. Atheneum, New York (1967)
36. Wikipedia. The united states privacy act of 1974 – Wikipedia, the Free Encyclopedia (1974), http://en.wikipedia.org/wiki/Privacy_Act_of_1974 (accessed May 12, 2009)
37. Wikipedia. Privacy — Wikipedia, the Free Encyclopedia (2009),
    http://en.wikipedia.org/wiki/Privacy#cite_note-0 (accessed May 12, 2009)
38. Yiu, M.L., Jensen, C.S., Huang, X., Lu, H.: Spacetwist: Managing the trade-offs among location privacy, query performance, and query accuracy in mobile services. In: International Conference on Data Engineering, pp. 366–375 (2008)
39. You, T.-H., Peng, W.-C., Lee, W.-C.: Protecting moving trajectories with dummies. In: IEEE International Conference on Mobile Data Management, pp. 278–282 (2007)

Querying and Cleaning Uncertain Data

Reynold Cheng

Department of Computer Science, The University of Hong Kong
Pokfulam Road, Hong Kong
ckcheng@cs.hku.hk
http://www.cs.hku.hk/~ckcheng

Abstract. The management of uncertainty in large databases has recently attracted tremendous research interest. Data uncertainty is inherent in many emerging and important applications, including location-based services, wireless sensor networks, biometric and biological databases, and data stream applications. In these systems, it is important to manage data uncertainty carefully, in order to make correct decisions and provide high-quality services to users. To enable the development of these applications, uncertain database systems have been proposed. They consider data uncertainty as a "first-class citizen", and use generic data models to capture uncertainty, as well as provide query operators that return answers with statistical confidences.

We summarize our work on uncertain databases in recent years. We explain how data uncertainty can be modeled, and present a classification of probabilistic queries (e.g., range query and nearest-neighbor query). We further study how probabilistic queries can be efficiently evaluated and indexed. We also highlight the issue of removing uncertainty under a stringent cleaning budget, with an attempt of generating high-quality probabilistic answers.

Keywords: uncertain databases, probabilistic queries, quality management.

1 Introduction

Data uncertainty is an inherent property in a number of important and emerging applications. Consider, for example, a habitat monitoring system used in scientific applications, where data such as temperature, humidity, and wind speed are acquired from a sensor network. Due to physical imperfection of the sensor hardware, the data obtained in these environments are often inaccurate [14]. Moreover, a sensor cannot report its value at every point in time, and so the system can only obtain data samples at discrete time instants. As another example, in the Global-Positioning System (GPS), the location collected from the GPS-enabled devices (e.g., PDAs) can also be contaminated with measurement and sampling error [26,15]. These location data, which are transmitted to the service provider, may further encounter some amount of network delay. In biometric databases, the attribute values of the feature vectors stored are not exact [2].

K. Rothermel et al. (Eds.): QuaCon 2009, LNCS 5786, pp. 41–52, 2009.
© Springer-Verlag Berlin Heidelberg 2009

Hence, the data collected in these applications are often imprecise, inaccurate, and stale.

Services or queries that base their decisions on these data can produce erroneous results. There is thus a need to manage these data errors more carefully. In particular, the idea of *probabilistic query* (PQ in short), which is a kind of queries that handle data uncertainty, has been recently proposed. The main idea of a PQ is to consider the models of the data uncertainty (instead of just the data value reported), and augment probabilistic guarantees to the query results. For example, a traditional nearest neighbor query asking who is the closest to a given point q can tell the user that John is the answer, while a PQ informs the user that John has a probability of 0.8 of being the closest to q. These probabilities reflect the degree of *ambiguity* of query results, thereby facilitating the system to produce a more confident decision. In this paper, we will examine different types of PQs, as well as techniques for improving their evaluation.

Another interesting issue is about the interpretation of probability values in the query answers. In general, a query answer can consist of numerous probability values, making it hard for a user to interpret the likelihood of their answers. A *quality metric* is thus desired, which computes a real-valued score for a probabilistic query answer [8,17]. This metric serves as a convenient indicator for the user to understand how vague his/her answer is, without the need of interpreting all the probabilities present in the answer. For example, if the score of his/her query answer is high, the user can immediately understand that the quality of his/her answer is good. We discuss how such quality metrics can be used to guide the removal of uncertainty from a database, with the goal of optimizing the quality of a probabilistic query result.

The rest of the paper is described as follows. We first examine one of the important models of uncertainty in Section 2. Then, the issues of classifying, evaluating and indexing important PQs in a large database will be addressed in Section 3. We examine the issues of cleaning uncertain data for the purpose of improving query answer quality, in Section 4. In Section 5, we conclude the paper and discuss interesting directions.

2   Modeling Uncertain Data

To understand a PQ, let us first discuss a commonly-used model of data uncertainty. This model, called *attribute uncertainty model*, assumes that the actual data value is located within a closed region, called the *uncertainty region*. In this region, a non-zero probability density function (*pdf*) of the value is defined, where the integration of pdf inside the region is equal to one. The cumulative density function (*cdf*) of the item is also provided. In an LBS, a normalized Gaussian pdf is used to model the measurement error of a location stored in a database [26,15] (Figure 1(a)). The uncertainty region is a circular area, with a radius called the "distance threshold"; the newest location is reported to the system when it deviates from the old one by more than this threshold (Figure 1(a)). Gaussian distributions are also used to model values of a feature vector in biometric databases [2]. Figure 1(b) shows the histogram of temperature values in

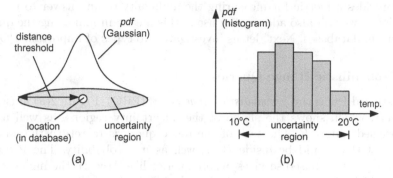

Fig. 1. Location and sensor uncertainty

a geographical area observed in a week. The pdf, represented as a histogram, is an arbitrary distribution between 10^oC and 20^oC.

A logical formulation of queries for the attribute uncertainty model has been recently studied in [21,25]. Other variants have also been proposed. In [18], piecewise linear functions are used to approximate the cdf of an uncertain item. Sometimes, point samples are derived from an item's pdf [16,22]. In the *existential uncertainty model*, every object is represented by the value in the space, as well as the probability that this object exists [12]. Some other kinds of models, e.g., tuple uncertainty, are addressed in [13,14,19,1]. A detailed survey of uncertainty models is presented in [23].

3   Probabilistic Queries

In this section, we explore the details of probabilistic queries, which are evaluated upon uncertain data. We first develop a query classification scheme in Section 3.1. Then, we address the efficient evaluation of two important probabilistic queries, namely range queries and nearest-neighbor queries, in Sections 3.2 and 3.3.

3.1   Query Classification

Given the uncertainty model, the semantics of PQs can be defined. We proposed a classification scheme for different types of PQ [8]. In that scheme, a PQ is classified according to the forms of answers. An *entity-based query* is one that returns a set of objects (e.g., list of objects that satisfy a range query or join conditions), whereas a *value-based query* returns a single numeric value (e.g., value of a particular sensor). Another criterion is based on whether an *aggregate* operator is used to produce results. An aggregate query is one where there is interplay between objects that determines the results (e.g., a nearest-neighbor query). Based on these two criteria, four different types of probabilistic queries are defined. Each query type has its own methods for computing answer probabilities. In [8], the notion of *quality* has also been defined for each query type, which provides a metric for measuring the ambiguity of an answer to the PQ. In Section 4, we will also address the use of this metric in managing the quality of uncertain databases. Next, let us investigate two types of important PQs.

3.2 Probabilistic Range Queries

A well-studied PQ is the *probabilistic range query* (PRQ). Figure 2(a) illustrates this query, which shows the shape of the uncertainty regions, as well as the user-specified range, R. The task of the range query is to return the name of each object that could be inside R, as well as its probability. The PRQ can be used in location-based services, where queries like: "return the names of the suspect vehicles in a crime scene" can be asked. It is also used in sensor network monitoring, where sensor IDs whose physical values (e.g., temperature, humidity) are returned to the user. Figure 2(a) shows the probability values of the items (A, B, and C) that are located inside R. A PRQ is an *entity-based* query, since it returns a list of objects. It is also a *non-aggregate* query, because the probability of each object is independent of the existence of other objects [8].

To compute an item's probability for satisfying the PRQ, one can first find out the overlapping area of each item's region within R (shaded in Figure 2(a)), and perform an integration of the item's pdf inside the overlapping area. Unfortunately, this solution may not be very efficient, since expensive numerical integration may need to be performed if the the item's pdf is arbitrary [11]. Even if an R-tree is used to prune items that do not overlap R, the probability of each item which is non-zero still needs to computed. A more efficient solution was developed in [11], where the authors proposed a user-defined constraint, called the *probability threshold P*, with $P \in (0,1]$. An item is only returned if its probability of satisfying the PRQ is not less than P. In Figure 2(a), if $P = 0.6$, then only A and C will be returned to the user. Under this new requirement, it is possible to incorporate the uncertainty information of items into a spatial index (such as R-tree). The main idea is to precompute the *p-bounds* of an item. A p-bound of an uncertain item is essentially a function of p, where $p \in [0, 0.5]$. In a 2D space, it is composed of four line segments, as illustrated by the hatched region in Figure 2(b). The requirement of right p-bound (illustrated by the thick solid line) is that the probability of the location of the item on the right of the line has to be exactly equal to p (the shaded area). Similarly, the probability of the item on the left of the left p-bound is exactly equal to p. The remaining line segments (top and bottom p-bounds) are defined analogously. Once these p-bounds are known, it is possible to know immediately whether an item satisfies the PRQ. Figure 2(b) shows that a range query R overlaps the item's uncertainty region, but does not cut the right p-bound, where p is less than P. Since the integration of the item's pdf inside the overlapping area of R and the uncertainty region cannot be larger than P, the item is pruned without doing the actual probability computation.

By precomputing a finite number of p-bounds, it is possible to store them in a modified version of the R-tree. Called *Probability Threshold Index* (PTI), this index can facilitate the pruning of uncertain items in the index level [11].

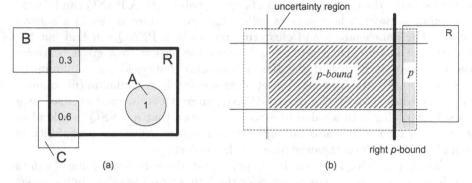

Fig. 2. Probabilistic range queries over uncertain items, showing (a) the probability of each item, and (b) the p-bound of an uncertain item

Compared with the R-tree which uses the MBR of the object for pruning, the use of p-bounds in the PTI provides more pruning power. In Figure 2(b), for example, although the range R overlaps with the MBR of the object (dashed rectangle), it does not cut the p-bounds, and so it can be pruned by the PTI but not by the R-tree. [11] also examined special cases of pdf (uniform and Gaussian distributions) and proposed an indexing scheme where p-bounds can be computed on-the-fly without being stored in the index. Since storing p-bounds for high-dimensional uncertain items can be expensive, Tao et al. [27,28] proposed a variant of PTI called *U-tree*, which only stores approximate information of p-bounds in the index. With these improvements, it is possible to index uncertain items in the high-dimensional space. The p-bound techniques were also used in [10] to facilitate the processing of join queries over uncertain spatial data.

Another PRQ evaluation technique was recently proposed by Ljosa et al. [18], who used piecewise linear functions to approximate the cdf of an uncertain item in order to avoid expensive integration. They also described an index that stored these piecewise linear functions, so that a PRQ can be evaluated more efficiently. More recently, the problem of evaluating *imprecise location dependent range queries* is studied [3,28]. This is a variant of PRQ, where the range query is defined with reference to the (imprecise) position of the query issuer. For instance, if the query issuer looks for his friends within 2 miles of his current position, and his position is uncertain, then the actual query range (a circle with a 2-mile radius) cannot be known precisely. In [3], we proposed several approaches to handle these queries, by (1) using the Minkowski Sum (a computational geometry technique), (2) switching the role of query issuer and data being queried, and (3) using p-bounds. Using detailed experiments on real datasets, we showed that these methods allow this kind of query to be efficiently evaluated.

3.3   Nearest-Neighbor Queries

Another important PQ for uncertain items is the probabilistic nearest-neighbor queries (PNNQ in short). This query returns the non-zero probability of each object for being the nearest neighbor of a given point q [8]. A PNNQ can be used in a sensor network, where sensors collect the temperature values in a natural habitat. For data analysis and clustering purposes, a PNNQ can find out the district(s) whose temperature values is (are) the closest to a given centroid. Another example is to find the IDs of sensor(s) that yield the minimum or maximum wind-speed from a given set of sensors [8,14]. A minimum (maximum) query is essentially a special case of PNNQ, since it can be characterized as a PNNQ by setting q to a value of $-\infty$ (∞). Notice that a PNNQ is a kind of *entity-based* query. It is also *non-aggregate* in nature, since the probability of each object *depends* on the existence of other objects [8].

Evaluating a PNNQ is not trivial. In particular, since the exact value of a data item is not known, one needs to consider the item's possible values in its uncertainty region. Moreover, since the PNNQ is an entity-based aggregate query [8], an item's probability depends not just on its own value, but also on the relative values of other objects. If the uncertainty regions of the objects overlap, then their pdfs must be considered in order to derive their corresponding probabilities. This is unlike the evaluation of PRQ, where each item's probability can be computed independent of others. To evaluate PNNQ, one method is to derive the pdf and cdf of each item's distance from q. The probability of an item for satisfying the PNNQ is then computed by integrating over a function of distance pdfs and cdfs [8,9,14]. In [9], an R-tree-based solution for PNNQ was presented. The main idea is to prune items with zero probabilities, using the fact that these items' uncertainty regions must not overlap with that of an item whose maximum distance from q is the minimum in the database. The *probabilistic verifiers*, recently proposed in [5], are algorithms for efficiently computing the lower and upper bounds of each object's probability for satisfying a PNNQ. These algorithms, when used together with the probability threshold defined by the user, avoid the exact probability values to be calculated. In this way, a PNNQ can be evaluated more efficiently. Recently, in [7], we extended the concept of probabilistic verifiers to efficiently handle probabilistic k-nearest-neighbor queries (i.e., finding the probability that k objects together are closest to q).

There are two other solutions for PNNQ that base on a different representation of uncertain items. Kriegel et al. [16] used the Monte-Carlo method, where the pdf of each object was sampled as a set of points. The probability was evaluated by considering the portion of points that could be the nearest neighbor. In [18], Ljosa et al. used piecewise linear representation of the cdf for an uncertain item to propose efficient evaluation and indexing techniques.

We conclude this section with another important entity-based aggregate query over uncertain items, namely the probabilistic skyline queries [22]. A skyline query returns a set of items that are not dominated by other items in all dimensions. In that paper, the issues of defining and computing the probability that an uncertain item was in the skyline were addressed. Two bounding-pruning-refining based algorithms were developed: the bottom-up algorithm used selected instances of uncertain items to prune other instances of uncertain items, while the top-down algorithm recursively partitions the instances of uncertain items into subsets. The authors showed that both techniques enable probabilistic sky-line queries to be efficiently computed.

4   Cleaning Uncertain Data

In this section, we explain how the probability values of the answers to PQs can be used to improve the quality of queries or services provided to the users. Section 4.1 discusses the issues in the cleaning of a database for attribute uncertainty. Section 4.2 outlines the cleaning process for the probabilistic database model.

4.1   Attribute Uncertainty

As discussed before, probability values in a query answer reflect the *ambiguity* of a query result (which are due to impreciseness of the data being evaluated). Consider Figure 3, where a monitoring server maintains the pdf of the temperature values acquired from four wireless sensors $(T_1, \ldots, T_4)$. Let us further suppose that a probabilistic range query (with a specified range $[10^oC, 20^oC]$) is issued on these four sensor data values, and produce the following answer:

$$\{(T_1, 0.9), (T_2, 0.5)\}$$

Since T_1 has a chance of 0.9 for satisfying the query, we know that T_1 is very likely to be located inside $[10^oC, 20^oC]$. The case of T_2 is more vague: it could either be inside or outside the specified range, with equal chances (0.5).

Quality Metric. In general, a query answer may consists of numerous probability values, making it hard for a query user to interpret the likelihood of their answers. A *quality metric* is thus desired, whose purpose is to compute a real-valued score for a probabilistic query answer [8,17]. This serves as a convenient

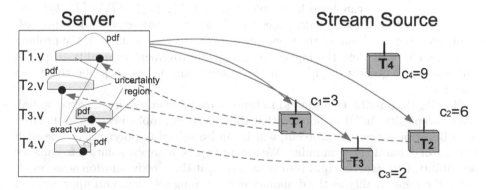

Fig. 3. Probing of Sensor Data for Uncertainty Reduction

indicator for the user to understand how vague his answer is, without the need of interpreting all probabilities present in the answer – e.g., if the score of his query answer is high, the user can immediately know that the quality of his answer is good.In [4], we defined a quality score for a probabilistic range query based on the definition of entropy [24]. This metric quantifies the degree of query answer uncertainty by measuring the amount of information presented in a query.

Budget-Limited Uncertainty Removal. A quality score enables us to address the question: "how can the quality of my query answer be improved?" Consider Figure 3 again. Suppose that the sensors have not reported their values for a long time. As a result, the sensor data kept in the server have a large degree of uncertainty. Consequently, the query answer quality is low (i.e., the query answers are vague), and a user may request the server to give him/her an answer with a higher quality. To satisfy the user's request, the system can acquire (or **probe**) the current values from the sensors, in order to obtain more precise information (i.e., possibly with a smaller uncertainty interval). A higher quality score for the query user's answer can then potentially be attained. In fact, if all the items $(T_1, \ldots, T_4)$ are probed, then the server will have up-to-date knowledge about external environments, thereby achieving the highest query quality.

In reality, it is unlikely that a system can always maintain an accurate state of the external environment, since probing a data item requires precious resources (e.g., wireless network bandwidth and batter power). It is thus not possible for the system to probe the data from all the sources in order to improve the quality of a query request. A more feasible assumption is that the system assigns to the user a certain amount of "resource budget", which limits the maximum amount of resources invested for a particular query. The question then becomes "how can the quality of a probabilistic query be maximized with probing under tight resource constraints?" In Figure 3, $c_1, \ldots, c_4$ are the respective costs for probing $T_1, \ldots, T_4$. The cost value of each sensor may represent the number of hops required to receive a data value from the sensor. Let us also assume that a query is associated with a resource budget of 8 units. If we wish to improve the quality for this query, we can have five *probing sets*: $\{T_1\}$, $\{T_2\}$, $\{T_3\}$, $\{T_1, T_2\}$ and $\{T_2, T_3\}$. Each of these sets describes the identities of the sensors to be probed. Moreover, the total sum of their probing costs is less than 8 units. If the probing of T_2 and T_3 will yield the highest degree of improvement in quality (e.g., the entropy-based quality in [4]), then the system only needs to probe these two sensors.

Testing the possible candidates in a brute-force manner requires an exponential-time complexity. In [4] we showed that this selection problem can be formulated as a linear optimization problem, which can be solved in polynomial time with the use of dynamic programming. We also presented a greedy solution to enhance scalability. Our experimental results showed that the greedy solution achieves almost the same quality as the dynamic-programming solution. Our approach can generally be applied to any multi-dimensional uncertain data, where the pdf's are arbitrary. We further considered the scenario where a group of query users share

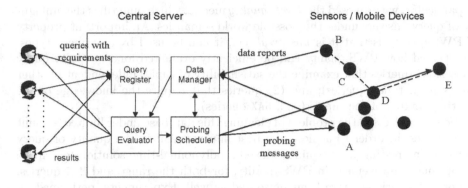

Fig. 4. System Architecture for Uncertain Data Cleaning

the same resource budget. This represents the case when a system allocates its resources to users with the same priority. We extended our basic solution (tailored for a single query) to address this problem.

System Architecture. Figure 4 describes the architecture of the system used in this paper. The *Data Manager* caches the value ranges and corresponding pdf of remote sensors. The *Query Register* receives queries from the users. The *Query Evaluator* evaluates the queries based on the information stored in the *Data Manager*. The *Probing Scheduler* is responsible for generating a *probing set* for each query – essentially the set of sensors to be probed. The benefits and costs of probing actions will be taken into account by the *Probing Scheduler* in deciding the what sensors to be consulted. More specifically, a probabilistic query is handled in four major steps:

- Step 1. The query is evaluated by the *Query Evaluator* based on the data cached in the *Data Manager*.
- Step 2. The *Probing Scheduler* decides the content of probing set.
- Step 3. The *Probing Scheduler* sends probing commands to the sensors defined in the probing set.
- Step 4. The *Query Evaluator* reevaluates the query based on the refreshed data returned to the *Data Manager*, and returns results to the query issuer.

Using the above steps, answers for a probabilistic query can be returned to a user with the best quality guaranteed under a limited cleaning budget.

4.2   Probabilistic Database

We have also considered the problem of quality management in *probabilistic databases*, which augments each tuple with a probability. Each tuple belongs to only one *x-tuple*, which is essentially a distribution of tuples. This database can be used to store uncertain data like those obtained from the integration of different web sources. In [6], we examine the quality notion for this kind of databases.

In particular, we proposed the *PWS-quality* metric, which quantifies the ambiguity of query answers under the possible world semantics. An important property of PWS-quality is that it is *universal*, i.e., it can be used by any query types. We studied how PWS-quality can be efficiently evaluated for two major query classes: (1) queries that examine the satisfiability of tuples independent of other tuples (e.g., range queries); and (2) queries that require the knowledge of the relative ranking of the tuples (e.g., MAX queries).

We also addressed the problem of cleaning this database under limited amount of resources, in order to achieve the best improvement in the quality of query answers. Interestingly, we again achieved a polynomial-time solution to achieve an optimal improvement in PWS-quality, for both the range and MAX queries. Other fast heuristics have been presented as well. Experiments, performed on both real and synthetic datasets, show that the PWS-quality metric can be evaluated quickly, and that our cleaning algorithm provides an optimal solution with high efficiency. To our best knowledge, this is the first work that develops a quality metric for a probabilistic database, and investigates how such a metric can be used for data cleaning purposes. Readers can refer to [6] for more details.

5   Conclusions and Future Work

Uncertainty management is an important issue in emerging applications like location-based services, road traffic monitoring, wireless sensor network applications, and biometric feature matching, where the data collected from the physical environments cannot be obtained with a full accuracy. Recent works have proposed to inject uncertainty to a user's location for location privacy protection [3]. The issue of managing uncertainty in a large database has also attracted plenty of research attentions; particularly, uncertain database prototypes have been developed (e.g., [20,13,19,1]). In this paper, we reviewed three important aspects of uncertainty management: (1) modeling of uncertainty; (2) classification and evaluation of probabilistic queries; and (3) uncertain data cleaning and improvement of probabilistic results.

A lot of work remains to be done in this area. An important future work will be the definition and efficient evaluation of important queries, such as reverse nearest-neighbor queries. Besides, we would like to investigate how to clean uncertain data for other kinds of queries and uncertainty models. It will also be interesting to study the development of data mining algorithms for uncertain data. Another direction is to study spatio-temporal queries over historical spatial data (e.g., trajectories of moving objects). It is also interesting to study the evaluation of continuous queries queries that reside in the system for an extensive amount of time and are commonly found in location-based services. Other works include revisiting query cost estimation and user interface design that allows users to visualize uncertain data. A long term goal is to consolidate these research ideas and develop a comprehensive spatio-temporal database system with uncertainty management facilities.

Acknowledgments

Reynold Cheng was supported by the Research Grants Council of Hong Kong (Projects HKU 513307, HKU 513508), and the Seed Funding Programme of the University of Hong Kong (grant no. 200808159002).

References

1. Antova, L., Koch, C., Olteanu, D.: Query language support for incomplete information in the maybms system. In: Proc. VLDB (2007)
2. Böhm, C., Pryakhin, A., Schubert, M.: The gauss-tree: Efficient object identification in databases of probabilistic feature vectors. In: Proc. ICDE (2006)
3. Chen, J., Cheng, R.: Efficient evaluation of imprecise location-dependent queries. In: Proc. ICDE (2007)
4. Chen, J., Cheng, R.: Quality-aware probing of uncertain data with resource constraints. In: Ludäscher, B., Mamoulis, N. (eds.) SSDBM 2008. LNCS, vol. 5069, pp. 491–508. Springer, Heidelberg (2008)
5. Cheng, R., Chen, J., Mokbel, M., Chow, C.: Probabilistic verifiers: Evaluating constrained nearest-neighbor queries over uncertain data. In: Proc. ICDE (2008)
6. Cheng, R., Chen, J., Xie, X.: Cleaning uncertain data with quality guarantees. In: Proc. VLDB (2008)
7. Cheng, R., Chen, L., Chen, J., Xie, X.: Evaluating probability threshold k-nearest-neighbor queries over uncertain data. In: Proc. EDBT (2009)
8. Cheng, R., Kalashnikov, D., Prabhakar, S.: Evaluating probabilistic queries over imprecise data. In: Proc. ACM SIGMOD, pp. 551–562 (2003)
9. Cheng, R., Kalashnikov, D.V., Prabhakar, S.: Querying imprecise data in moving object environments. IEEE TKDE 16(9) (September 2004)
10. Cheng, R., Singh, S., Prabhakar, S., Shah, R., Vitter, J., Xia, Y.: Efficient join processing over uncertain data. In: Proc. CIKM (2006)
11. Cheng, R., Xia, Y., Prabhakar, S., Shah, R., Vitter, J.S.: Efficient indexing methods for probabilistic threshold queries over uncertain data. In: Proc. VLDB, pp. 876–887 (2004)
12. Dai, X., Yiu, M.L., Mamoulis, N., Tao, Y., Vaitis, M.: Probabilistic spatial queries on existentially uncertain data. In: Proc. SSTD, pp. 400–417 (2005)
13. Dalvi, N., Suciu, D.: Efficient query evaluation on probabilistic databases. In: VLDB (2004)
14. Deshpande, A., Guestrin, C., Madden, S., Hellerstein, J., Hong, W.: Model-driven data acquisition in sensor networks. In: Proc. VLDB (2004)
15. Pfoser, D., Jensen, C.: Capturing the uncertainty of moving-objects representations. In: Proc. SSDBM (1999)
16. Kriegel, H., Kunath, P., Renz, M.: Probabilistic nearest-neighbor query on uncertain objects. In: Kotagiri, R., Radha Krishna, P., Mohania, M., Nantajeewarawat, E. (eds.) DASFAA 2007. LNCS, vol. 4443, pp. 337–348. Springer, Heidelberg (2007)
17. Lazaridis, I., Mehrotra, S.: Approximate selection queries over imprecise data. In: ICDE (2004)
18. Ljosa, V., Singh, A.: Apla: Indexing arbitrary probability distributions. In: Proc. ICDE, pp. 946–955 (2007)
19. Mar, O., Sarma, A., Halevy, A., Widom, J.: ULDBs: databases with uncertainty and lineage. In: VLDB (2006)

20. Mayfield, C., Singh, S., Cheng, R., Prabhakar, S.: Orion: A database system for managing uncertain data, ver. 0.1 (2006), http://orion.cs.purdue.edu
21. Parker, A., Subrahmanian, V., Grant, J.: A logical formulation of probabilistic spatial databases. IEEE TKDE 19(11) (2007)
22. Pei, J., Jiang, B., Lin, X., Yuan, Y.: Probabilistic skylines on uncertain data. In: Proc. VLDB (2007)
23. Sarma, A., Benjelloun, O., Halevy, A., Widom, J.: Working models for uncertain data. In: Proc. ICDE (2006)
24. Shannon, C.: The Mathematical Theory of Communication. University of Illinois Press, Urbana (1949)
25. Singh, S., Mayfield, C., Shah, R., Prabhakar, S., Hambrusch, S., Neville, J., Cheng, R.: Database support for probabilistic attributes and tuples. In: Proc. ICDE (2008)
26. Sistla, P.A., Wolfson, O., Chamberlain, S., Dao, S.: Querying the uncertain position of moving objects. In: Etzion, O., Jajodia, S., Sripada, S. (eds.) Dagstuhl Seminar 1997. LNCS, vol. 1399, Springer, Heidelberg (1998)
27. Tao, Y., Cheng, R., Xiao, X., Ngai, W.K., Kao, B., Prabhakar, S.: Indexing multi-dimensional uncertain data with arbitrary probability density functions. In: Proc. VLDB, pp. 922–933 (2005)
28. Tao, Y., Xiao, X., Cheng, R.: Range search on multidimensional uncertain data. ACM TODS 32(3) (2007)

Spatial Embedding and Spatial Context

Christopher Gold

Department of Computing and Mathematics, University of Glamorgan, Wales, UK
and Department of Geoinformatics, Universiti Teknologi Malaysia (UTM)
chris.gold@gmail.com

Abstract. A serious issue in urban 2D remote sensing is that even if you can identify linear features it is often difficult to combine these to form the object you want – the building. The classical example is of trees overhanging walls and roofs: it is often difficult to join the linear pieces together. For robot navigation, surface interpolation, GIS polygon "topology", etc., isolated 0D or 1D elements in 2D space are incomplete: they need to be fully embedded in 2D space in order to have a usable spatial context. We embed all our 0D and 1D entities in 2D space by means of the Voronoi diagram, giving a space-filling environment where spatial adjacency queries are straightforward. This has been an extremely difficult algorithmic problem. We show recent results. If we really want to move from exterior form to building functionality we must work with volumetric entities (rooms) embedded in 3D space. We thus need an adjacency model for 3D space, allowing queries concerning adjacency, access, etc. to be handled directly from the data structure, exactly as described for 2D space. We will show our recent results to handle this problem. We claim that an appropriate adjacency model greatly simplifies questions of spatial context of elements (such as walls) that may be extracted from raw data, allowing direct assembly of compound entities such as buildings. Relationships between compound objects provide solutions to building adjacency, robot navigation and related problems. If the spatial context can be stated clearly then other contextual issues may be greatly simplified.

1 Introduction

In this paper we claim that a spatial model of the real world can be represented as a graph, that the spatial relationships between the nodes of this graph are expressed in its dual, that these spatial relationships can be considered to be the spatial context, that this is true in 3D as well as 2D, and that we have an operational spatial data structure to prove it.

Let us first look at the small print. The real world we are thinking of is mostly "urban", with a relatively simple boundary structure. The spatial model consists of a connected collection of entities, both atomic and compound.

Compound entities (such as houses) are composed of sets of atomic entities (walls, roof planes, edges, vertices). A graph consists of edges and nodes: nodes are any type of atomic entity; edges show the connections between pairs of atomic entities – usually representing adjacency. The dual graph consists of a set of elements each of

K. Rothermel et al. (Eds.): QuaCon 2009, LNCS 5786, pp. 53–64, 2009.

which is the dual of a node or edge in the primal graph. In 2D the dual of a node is a region – a polygon, and the dual of a region is a node. The dual of an edge is another "intersecting" edge, and vice-versa. In 3D the dual of a node is a volume and the dual of a volume is a node. The dual of an edge is a face, and vice versa.

Thus a primal node in 2D has a bounding area formed by dual edges. In 3D it has a bounding volume formed by dual faces – each of these faces is paired with a primal edge. These bounding edges or faces are common to two adjacent nodes, and thus show the adjacency relationships. This could be called the spatial context, as it describes the "world" around the node.

1.1 Context and Adjacency Layers

Our initial spatial model consists of a set of individual entities – e.g. data points or line segments. We will call this the "Entity Layer".

The "extent" of an entity defines its "context" – its set of neighbours embedded in the model space. Metric proximity is the usual definition of extent (the Voronoi diagram) but not the only one. This is the "Context Layer".

The boundaries separating pairs of extents define adjacent pairs of entities in 2D. These are stored in the dual "Adjacency Layer" (often the Delaunay triangulation).

Both the Context and Adjacency Layers may be treated as planar graphs of nodes, edges and regions. While at first sight both are not needed, as one may be derived from the other, they express different things, and sometimes both are needed simultaneously to describe some spatial feature. Two subsets of these that are of particular use in GIS are the crust and skeleton, and the buffer zone.

1.2 Additional Features: Crust, Skeleton and Buffer Zone

The skeleton/medial axis is a subset of the boundaries of the Context Layer, separating those entities that form separate "groups". (These groups need not be fully separate, as in inter-fingering rivers.) The separation may be proximal, using the Voronoi diagram, or by attributes, such as point labels. The crust (the dual of the remaining Context layer boundaries) connects portions of the Entity Layer that are not separated by the skeleton [1], [2].

Buffer zones partition the extents of a set of entities into those portions within a given distance of the generating entity, and those portions further away [3].

2 2D Spatial Context in GIS

In most GIS there are four major categories of features and related data structures: discrete objects; networks; polygonal maps; and surfaces [4]. We will illustrate the relevance of the Context Layer, and sometimes the Adjacency Layer, with examples.

2.1 Discrete Objects

These are either polygonal entities representing features with areal extent, or points representing "small" features. In addition we consider the case of mobile points.

2.1.1 Points

In Fig. 1 the extent of each outcrop observation is based on the idea of "What is the closest observation to my current location?" See [5]. The context of each observation is its set of neighbours: it is either surrounded by neighbours of the same rock type, or else some neighbours are different. The labelled skeleton (heavy line) separates rocks of different types. In another application the generating points might be tree crowns of differing species. In a dense forest the extent would indicate spreading, and hence tree size or volume. Tree population density at a local level is the reciprocal of individual areas.

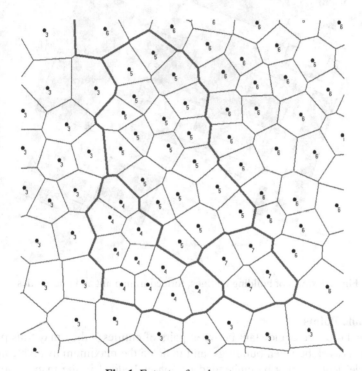

Fig. 1. Extents of rock outcrops

2.1.2 Discrete Polygons

A common example of discrete polygons is a map of building outlines (Fig. 2). Here the extent of each building indicates its neighbours (its context): the length of the common boundary may indicate the importance or relevance of the adjacency relationship [6]. (An alternate method is the area-stealing paradigm, as in Sibson interpolation [7].) As with the forestry example the area of extent provides a local measure of building density.

Buildings are complex entities, formed from atomic point and line entities. Each line segment has two halves, so the interior and exterior relationships of building portions may be examined.

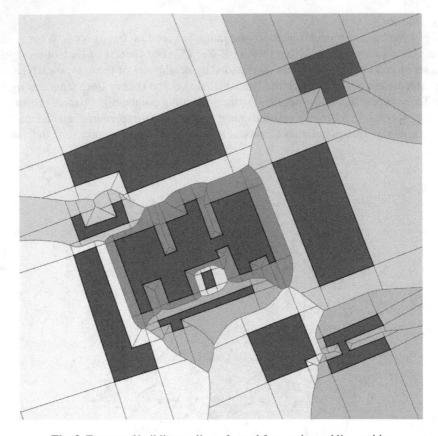

Fig. 2. Extents of building outlines, formed from point and line entities

2.1.3 Mobile Points

Extent boundaries are equidistant between pairs of entities – they may thus provide a navigation network between buildings, and indicate the maximum available clearance for a vehicle. Robots may be embedded or non-embedded in the map plane: if they are embedded then they may interact or collide with the other map entities: they behave like "boats". If they are not embedded they act like "planes" and do not interact with their environment. Embedded moving points are used for map construction, marine GIS etc. as the Voronoi diagram (context) is updated as it moves.

2.2 Networks

The most obvious examples of networks are roads and rivers. The first case allows closed loops, which may or may not be considered as regions. The second case assumes a tree structure and directionality of movement. Here the context layer gives not only the connectivity of the network (also valuable for detecting digitizing errors) but information about spatially close network segments as well.

2.2.1 Roads

Each half-line is separate, permitting representation of two-way traffic, and has its own extent. The result is a complete tessellation of the map space: this permits off-road queries of adjacency, as in finding the nearest road to some arbitrary off-road location – see Fig. 3, from [3].

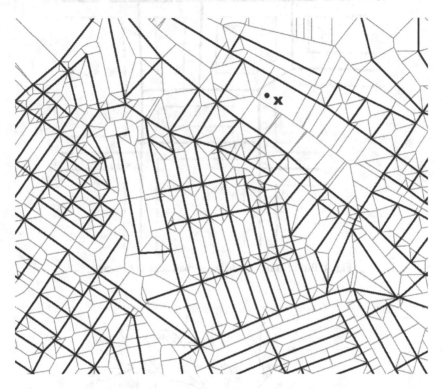

Fig. 3. A road network, showing extents of each segment

Fig. 4 (from [3]) gives a simple example of a buffer zone for part of a road network. Each point or line segment extent is split according to the buffer width and the pieces compiled together. This is the inverse of the normal procedure: here we first find the proximal regions and then query them with respect to the buffer width.

2.2.2 Rivers

River networks may be constructed of line segments, as for the roads. Alternatively, the hydrography may be digitized as closely spaced points and the Voronoi diagram built. Each cell is the context of one location on the river: it is either adjacent to the nearest upstream/downstream point, or else to points from an adjacent stream. The skeleton or medial axis, – the subset of Voronoi boundaries between widely separated points – forms a preliminary estimate of the watershed, while the dual of the remaining boundaries forms the crust, connecting the closely spaced points along the rivers (Fig. 5) [8].

Fig. 4. Buffer zones partitioning road segment extents

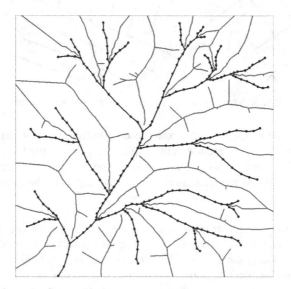

Fig. 5. Crust and skeleton of a simple digitized river network

Fig. 6 (from [8]) shows this in perspective view, with watershed heights based on Blum's height transform. The individual extents give the areas closest to each individual point, which reasonably estimate the rainfall reaching the river overland at that location. The crust, a subset of the dual Delaunay triangulation, gives the adjacency of the river segments. Summing the rainfall areas while following the rivers downstream gives an approximation of the total flow at any river location, as shown by the bars. Hence both primal and dual (context and adjacency) layers are needed for analysis.

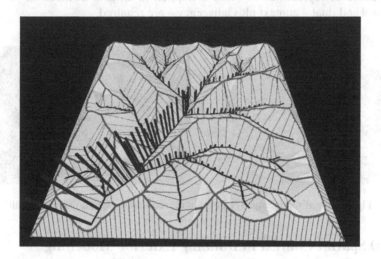

Fig. 6. Cumulative flow in a river network

2.2.3 Polygonal Maps

Standard choropleth maps, or maps of administrative districts, may be constructed as for road networks (Fig. 3). Labelling half-edge extents with the district label can be used to validate the digitizing process and to calculate areas.

2.3 Surfaces

Surfaces are fields, with "elevation" values that may be estimated at any location. This is true of any of the context models discussed so far, where the returned value may be label of the nearest entity. It may also be derived from the adjacency layer of a set of points, generating a traditional TIN model. More generally, a surface model consists of a set of data, a suitable spatial data structure, and an interpolation algorithm. The "area-stealing" or "natural neighbour" algorithm [7] is particularly stable, as the weighting function matches the neighbour selection procedure. It operates by embedding the query point in the map and calculating the relative areas stolen from the neighbouring data points, then using these to weight the data values. Thus the context of the query point is the basis for interpolation.

2.3.1 Runoff Models

Finite-difference runoff modelling is most often done using a grid of elevation estimates and then estimating flow between grid cells. This has a variety of problems (see [9]). Using a randomized set of query locations instead of a grid reduces the N-S, E-W bias. Fig. 7a shows a surface model with the extents of each elevation point (which is in the centre of each cell). These are the "buckets" that contain the water being transferred (rather than the usual square cells). Fig. 7b shows the adjacency graph for these elevation points – the Delaunay triangulation. (The heavy lines show the original contour data from which the elevation points were derived.) These graph edges between the elevation points give the gradients for flow between cells, taking the cell centre as representative. (This is implicit in all grid methods as well.) Thus both primal and dual – context plus adjacency – are required.

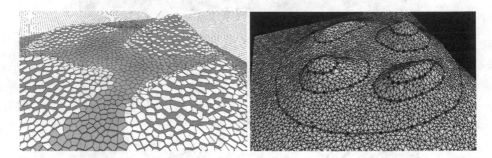

Fig. 7. Runoff modelling: a) extents of elevation points; b) adjacency graph

3   2.5D Spatial Context in Building Exterior Modelling

Let us look beyond the simple 2D primal and dual case. We will examine the urban situation: firstly for extruded building exteriors, and then for "functional" buildings with interior spaces.

"Context" was previously used to refer to the local arrangement of entities around the particular entity being described – points and line segments. We now start with point clouds (in this case from LIDAR) and must aggregate these individual points into usable higher level entities: surfaces (planes for our simplified model). We wish to return to our earlier definition of context for these entities, but first we must add some environmental knowledge in order to define our building planes. Various propositions can define atomic entities (planes) and objects (buildings): these should be as general as possible, to avoid restricting our model world unnecessarily. However, without some constraints we would be unable to recognize any entities beyond the initial points. We start with as simple propositions as possible [10].

Proposition 1: Buildings are collections of contiguous elevations that are higher than the surrounding terrain. Their boundaries are "walls". (We need to identify high areas surrounded by drops.)

Proposition 2: Walls have a specified minimum height, and this height difference is achieved within a very few "pixels". (These drops should be steep, and of some significant height.)

Proposition 3: A building consists of a high region entirely surrounded by walls. (To simplify detection of complete buildings, lean-tos and buildings against slopes will be excluded.) Fig. 8a shows the results of segmenting the LIDAR point cloud into "high" and "low" regions by cataloguing them into index cells – which are then split if necessary.

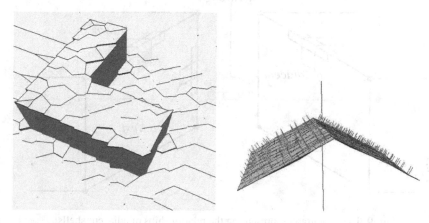

Fig. 8. Building reconstruction from LIDAR data: a) high/low point segregation; b) roof plane detection from vector normals

Proposition 4: Roofs are made up of planar segments, most of whose constituent triangles have similar vector normals. (We will limit ourselves to planar roof entities: these can be detected by standard TIN triangle orientation.) Fig. 8b illustrates a simple roof.

Proposition 5: The relationships between roof planes may be represented as a dual graph. (At this stage our atomic entities (nodes) are planar segments. Adjacency is determined by their intersections. The result is an adjacency graph.)

Proposition 6: Building exteriors, together with the adjacent terrain, form a portion of the global "Polyhedral Earth".

Proposition 7: Building interiors may be constructed as individual polyhedra, linked together and to the exterior by edges of the adjacency graph. (We now introduce volume entities, and our graph is fully 3D. This is described in the following section.)

4 3D Spatial Context in Volumetric Modelling

When we have volumes, as well as area, line and point entities, we need to work with the 3D duality properties. As mentioned in the Introduction, volume duals are points and face duals are edges. Thus each node entity in our graph may represent a point or a volume in the other space, and each edge may represent a face in the other space. Consequently our data structure remains a graph of nodes and edges. Adjacency of volumes is represented by an edge through their common face or faces. Fig. 9 [11] shows this for the Augmented Quad-Edge (AQE) of [12]. Each separate shell around a volume is constructed from the Quad-Edge (QE) of [13] – the change was that the original QE pointer to a 2D face was replaced by a pointer to the dual edge penetrating that face.

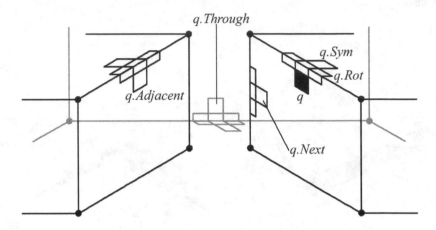

Fig. 9. The dual graph representing the relationships of adjacent shells

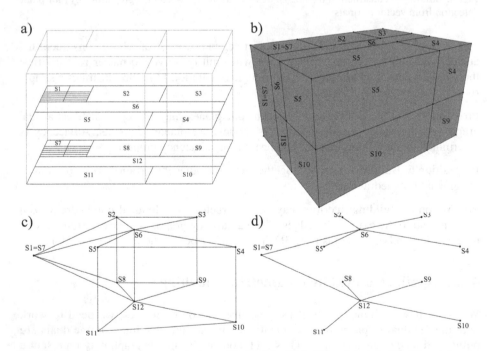

Fig. 10. Example structure representing a 3D model of a building interior. S1, S7 – staircase, S2-S5, S8-S11 rooms, S6, S12 – corridor: a) spatial schema, b) volumetric model of rooms, c) complete graph of connections between rooms, d) graph of accessible connections between rooms.

5 The Dual Half-Edge Data Structure

The Dual half-Edge (DHE) replaces the AQE by noting that the QE or AQE consists of a pair of primal half-edges (HEs) in the primal linked to a matching pair in the dual. These may be reconfigured as two pairs of matched half-edges, where each pair consists of one HE in each space. One of these may be considered to be the adjacency "layer" connecting the vertices of individual shells. The other is the context "layer", specifying the relationships between volumes.

Interestingly, the DHE can be shown to be a common ancestor of both the 2D QE and of the 3D structures. Two DHEs may be combined to give the primal/dual QE element in 2D. They may also be combined to give the "Cardboard and Tape" 3D CAD model of [14]. Recent work has shown that multi-shell, 3D Euler operators, with automatic construction of the dual, may be achieved from the same primitives. Fig. 10, from [14] shows a final example. A simple building is modelled by a set of boxes. The dual graph is constructed automatically, expressing the relationships between the rooms – and hence producing a system for escape route planning (the desired context). Lee [15] constructed a simpler model on a floor-by-floor basis. Thus in both 2D and 3D the dual graph, or "Context", is fundamental to understanding the spatial relationships between entities.

References

1. Blum, H.: A Transformation for Extracting New Descriptors of Shape, pp. 362–380. MIT Press, Cambridge (1967)
2. Gold, C.M., Snoeyink, J.: A One-Step and Skeleton Extraction Algorithm. Algorithmica, 144–163 (2001)
3. Dakowicz, M.: A Unified Spatial Data Structure for GIS. Ph.D. Thesis, University of Glamorgan, Wales, UK, 225p. (submitted, 2009)
4. Burrough, P.A., McDonnell, R.: Principles of Geographical Information Systems. Oxford University Press, New York (1998)
5. Okabe, A., Boots, B., Sugihara, K., Chiu, S.N.: Spatial Tessellations - Concepts and Applications of Voronoi Diagrams, 2nd edn., 671 p. John Wiley and Sons, Chichester (2000)
6. Gold, C.M., Dakowicz, M.: Kinetic Voronoi/Delaunay Drawing Tools. In: Proceedings, 3rd International Symposium on Voronoi Diagrams in Science and Engineering, pp. 76–84 (2006)
7. Sibson, R.: A Brief Description of Natural Neighbor Interpolation. In: Barnett, V. (ed.) Interpreting Multivariate Data, pp. 21–36. John Wiley & Sons, Chichester (1982)
8. Gold, C.M., Dakowicz, M.: The Crust and Skeleton - Applications in GIS. In: Proceedings, 2nd. International Symposium on Voronoi Diagrams in Science and Engineering, pp. 33–42 (2005)
9. Dakowicz, M., Gold, C.M.: Finite Difference Method Runoff Modelling Using Voronoi Cells. In: Proceedings, 5th. ISPRS Workshop on Dynamic and Multi-dimensional GIS (DMGIS 2007), Urumqi, China (2007)
10. Gold, C.M., Tse, R.O.C., Ledoux, H.: Building Reconstruction - Outside and In. In: 3D Geo Information Systems. Lecture Notes in Geoinformation and Cartography, pp. 355–369 (2006)

11. Boguslawski, P., Gold, C.M.: Euler Operators and Navigation of Multi-shell Building Models. In: 3D Geo Information Systems. Lecture Notes in Geoinformation and Cartography (submitted, 2009)
12. Ledoux, H., Gold, C.M.: Simultaneous Storage of Primal and Dual Three-dimensional Subdivisions. Computers, Environment and Urban Systems 31, 393–408 (2007)
13. Guibas, L.J., Stolfi, J.: Primitives for the Manipulation of General Subdivisions and the Computation of Voronoi Diagrams. ACM Transactions on Graphics 4, 74–123 (1985)
14. Boguslawski, P., Gold, C.M.: Construction Operators for Modelling 3D Objects and Dual Navigation Structures. Lectures Notes in Geoinformation and Cartography, pp. 47–59 (2009)
15. Lee, J.: A Three-Dimensional Navigable Data Model to Support Emergency Response. In: Microspatial Built Environments. Annals of the Association of American Geographers, vol. 97, pp. 512–529 (2007)

A Context Quality Model to Support Transparent Reasoning with Uncertain Context*

Susan McKeever, Juan Ye, Lorcan Coyle, and Simon Dobson

System Research Group, School of Computer Science and Informatics
UCD, Dublin, Ireland
susan.mckeever@ucd.ie

Abstract. Much research on context quality in context-aware systems divides into two strands: (1) the qualitative identification of quality measures and (2) the use of uncertain reasoning techniques. In this paper, we combine these two strands, exploring the problem of how to identify and propagate quality through the different context layers in order to support the context reasoning process. We present a generalised, structured context quality model that supports aggregation of quality from sensor up to situation level. Our model supports reasoning processes that explicitly aggregate context quality, by enabling the identification and quantification of appropriate quality parameters. We demonstrate the efficacy of our model using an experimental sensor data set, gaining a significant improvement in situation recognition for our voting based reasoning algorithm.

1 Introduction

The information used by context-aware systems to recognise different contexts is often imperfect. Sensor data is prone to noise, sensor failure and network disruptions. Users actions can contribute to degradation of information quality, such as the failure of users to carry their locator tags. Further uncertainty can be introduced in the reasoning process, such as the use of fuzzy functions to quantify vague context or difficulty in defining accurate inference rules [14]. Existing work in the area of context quality focuses on two main areas: (1) The *qualitative* identification of context quality parameters, often as part of a context modelling exercise, such as the work done by [4,5,6]; and (2) the *quantitative* use of reasoning techniques that incorporate context uncertainty such as Bayesian networks [10] and fuzzy logic [7].

The qualitative work provides a useful vocabulary for identifying and modelling context quality. However, such measures are usually associated with 'context', without specification of quality for each *layer* of context. Quality issues for low level sensor data are different from those at higher levels of context and

---

* This work is partially supported by Enterprise Ireland under grant number CFTD 2005 INF 217a, and by Science Foundation Ireland under grant numbers 07/CE/I1147 and 04/RPI/1544.

K. Rothermel et al. (Eds.): QuaCon 2009, LNCS 5786, pp. 65–75, 2009.

a context quality model must reflect this [9]. Also, the *aggregation* of quality across the layers must be addressed in order to produce a meaningful and useable indicator of context quality to applications. This aggregation will support reasoning schemes that can propagate uncertainty from sensor level upwards. For example, Dempster Shafer [13] or voting algorithms [2] for context reasoning can incorporate explicit quantification of uncertainty of context sources.

This paper presents a UML-based structured model of context quality for each layer of context. We also include an aggregation model that contains a general set of quality measures and their propagation across context layers. Designers of context-aware systems can use our combined models to (1) identify and model context quality issues and (2) to support the specification of quality aggregation. In particular, context-aware systems using transparent reasoning techniques that aggregate quality from sensors upward will benefit from our modelling approach. We demonstrate our work by generating quality parameters for an experimental dataset. We incorporate these quality parameters into a voting-based reasoning algorithm. Our results show that situation recognition is significantly improved with the inclusion of our modelled context quality than when quality is not used.

This remainder of this paper is organised as follows: Section 2 describes related work by other researchers; Section 3 details our structured quality and aggregation models and their relevance to context reasoning schemes; In Section 4, we demonstrate and critique our model with an experimental dataset. Finally, in Section 5, we conclude our work and define our future research direction.

2   Related Work

Previous work on modelling context quality provides various well documented parameters for context quality, such as context *confidence* [3,10,11] to indicate probability of correctness and *freshness* [1,3,4,8] to indicate the degradation of information over time. Lower level sensor quality measures such as *precision*, *accuracy* and *resolution* [1,4] are used to define sensor data issues. Such work provides useful semantics for exploring the nature of context quality issues. Other modelling approaches include placeholders for quality parameters within structured models of context. For example, Henricksen and Indulska's [6] Object Role Modelling context model associates context facts with zero or more quality parameters and associated metrics. Similarly, Gu *et al.* [5] describe a context model that includes a quality ontology with specific parameters and metrics. They include a set of commonly used parameters in their ontology. Both of these models use similar modelling constructs for quality. However, sensor and situation quality parameters are not separately identified.

The modelling approaches described do not model or aggregate quality parameters at each layer of context. We address this as follows: (1) We provide a structured (UML) extendable context quality model that includes quality parameters for sensor, abstracted context and situations, as illustrated in

Figures 1 and 2. (2) We provide an aggregation model that aggregates quality across each layer of context, as shown in Figure 3. To illustrate our work, we provide a demonstration of how context quality can be used to support the reasoning process, as explained in Section 4.

3 Our Approach to Modelling Context Quality

Context-aware applications are usually decoupled from the intricacies of lower-level sensor data, Instead, they liaise with higher-level human-understandable situations [14], such as a user 'preparing breakfast' at home. For such applications, a usable expression of quality at situation level will be useful, rather than having to interpret lower level measures such as sensor precision. We approach our context quality modelling work with the following principles in mind: (1) Quality issues at each layer of context (sensor, abstracted context, situation) are different [9], requiring parameters to be specified for each layer; (2) Known aggregations between quality parameters along the layers should be shown to assist in quantifying quality. However, prescription of explicit aggregation formulae is avoided to allow for system-specific calculations; (3) A model of context quality should be extendable to allow for system-specific requirements; (4) A standard modelling technique should be used to maximise use of the structured model. We incorporate these principles into our structured (Figures 1 and 2) and aggregation models (Figure 3). These models are general and flexible enough to suit different types of sensors and context and their associated quality issues. We use UML for our structured model.

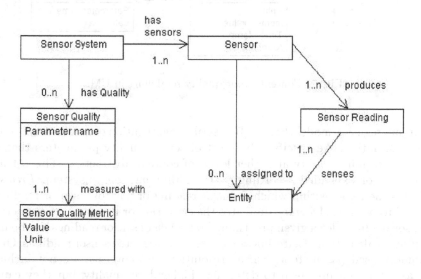

Fig. 1. Sensor quality modelling in UML

3.1   Modelling Sensor Quality

In our UML model, sensors represent any physical (e.g. Ubisense[1]) or virtual data source (e.g. tracked user calendar) that provides dynamic information about an entity, such as location of a user. Sensors are heterogeneous. Therefore, it is challenging to provide a set of quality parameters that applies across all sensors. Our structured model is generic and flexible, describing placeholders for modelling quality without prescribing what parameters to model. Our sensor quality model as shown in Figure 1 shows zero or more quality parameters for each sensor. For each quality parameter in our sensor class, one or more metrics may be stored. Each sensor system may have one or more sensors, and each sensor tracks one or more entities. A quality parameter is an instantiated sensor quality class and (at least one) sensor quality metric class. E.g., our structured model of our in-house Ubisense system includes a quality parameter of 'precisionx' (for x-axis), with a metric value of 1.65 and unit of metres.

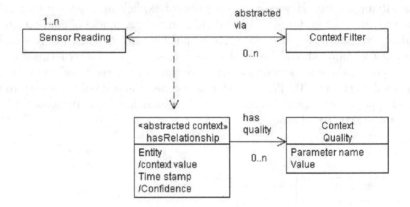

Fig. 2. Context event quality modelling in UML

Our aggregation model (Figure 3) describes aggregation of quality from sensor level to situation. We specify a base set of sensor quality parameters that are used to determine quality at higher levels of context abstraction. This set may be reduced or expanded according to the individual sensor system. *Precision* indicates the range within which a sensor reading or part of a sensor reading is believed to be true; *Accuracy* indicates the error rate or frequency of correctness of sensor readings, for a given precision; *Frequency* of sensor readings can be used to support calculation of a freshness measure of abstracted sensor readings. Other commonly used parameters such as resolution and coverage are not included here as we do not use them to determine higher level quality, but they can be instantiated in the structured model for system specific scenarios.

---

[1] Ubisense is a networked location system : www.ubisense.net

3.2   Modelling Abstracted Context Quality

As shown in Figure 2, sensor readings are abstracted to more meaningful context by passing one or more sensor readings through a static mapping or filter. E.g. a Ubisense coordinate of *(12.3, 32.4, 34.1, ID34)* may be mapped via a building map to 'John's desk'. Abstracted context is modelled as a UML association class that describes the relationship (e.g. has location, has temperature) between an entity class and a context filter class at a particular point in time. A context event is an instantiation of the association class for an instantiated entity class and context filter class for a particular time *t*. A context event may also be modelled as an association between two entities, such as 'John located near Susan'. Context confidence is included for each context event. It will be derived from zero or more context event quality parameters as used for the system in question. Our structured model therefore allows for any type of context relationship and associated quality to be modelled.

In our aggregation model (Figure 3), we capture context event quality in a base set of quality parameters; these quantify the imperfections of vague context (fuzzy membership), erroneous or conflicting context (reliability), imprecision (precise membership) and out of dateness (freshness). These values are combined to produce a final context event *confidence* value. As shown in Figure 3, context event confidence is a function of the other context event quality parameter values, such as the product of their values as per [2] or as their averaged value. *Fuzzy membership* is used when fuzzy context filters are used in the abstraction process. In our experimental work, a computer activity sensor that tracks keyboard/mouse activity classifies its context values as 'active' or 'inactive'. We apply a fuzzy membership value using a linear fuzzy membership function for the context filter. For context filters with crisp boundaries such as location, *precision membership* of the bounded value is derived using the sensor reading precision. This captures the impact of general sensor precision for a particular context instance. For example, we use Ubisense precision to define a bounded area within which the true Ubisense coordinate should occur, as described in [9]. This area may intersect more than one location (e.g. desk), leading to quantifiable membership of each desk location. E.g. User John is 0.8 in Desk 1, and 0.2 at Desk 2. *Reliability* captures the error rate associated with a context event. The reliability can incorporate any sources of error, including user error and sensor system accuracy. It can be measured objectively by training or observation. *Freshness* indicates the extent to which time has eroded the credibility of the context event. It can be calculated in various ways, such as use of a decay function [9], or a valid lifetime, derived from sensor reading timestamp or frequency. We avoid prescribing explicit hard-coded formulae for freshness and other parameters as their calculation may differ from one scenario to another.

3.3   Modelling Situation Quality

The provision of *situation confidence* allows adaptive applications to assess the risk associated with responding to a situation. Our structured model therefore

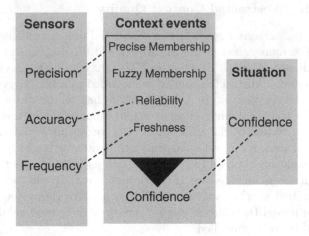

Fig. 3. Aggregation model: quality parameters and aggregations. The lines indicate how quality is aggregated; e.g accuracy at sensor level contributes to the reliability of a context event

explicitly includes at least one situation confidence parameter. Situation confidence is modelled as an attribute in a situation association class. The calculation of situation confidence will depend upon the reasoning scheme employed. For schemes that aggregate quality from sensor level upwards, situation confidence will be derived by fusing confidences of causal context events, as shown in our aggregation model (Figure 3).

3.4   Using Context Quality in the Reasoning Process

Reasoning with uncertain context typically involves reasoning schemes from the AI domain employed within context middleware, such as Bayesian networks [10], probabilistic logic [10], neural networks [11], Dempster Shafer [13], voting [2], and fuzzy logic [7]. The selection of one reasoning scheme over another will depend on a variety of factors such as: the availability of training data versus domain knowledge; the requirement for end users to understand the reasoning process; and the level of re-training required due to flux in the environment. Such reasoning schemes typically produce a quantified confidence in each of the possible situations. Applications can then use confidence to safeguard their adaptation strategy, such as accepting situations above a particular threshold value.

The extent to which sensor and context quality is *transparently* incorporated into the reasoning process depends on the reasoning mechanism involved. Bayesian networks can be naively trained using a 'black box' approach. In this way, there is no transparent quantification or aggregation of sensor or context event quality. The degree of uncertainty appears as 'output' via the root node variable's probabilities. Alternatively, other finer grained reasoning mechanisms can specifically aggregate lower level sensor and context quality in a way that is transparent to the systems designer. Examples of such mechanisms are voting [2], Dempster Shafer [13] and fuzzy logic [10]. Uncertain reasoning processes that transparently aggregate quality of context up to situation level *need to identify, model and aggregate appropriate quality parameters* along each layer of context. Our model supports the identification of quality issues and quality aggregation as part of such transparent reasoning processes.

4   Demonstration of Model

We used our model to identify and aggregate quality parameters for a sensed data set. We wanted to determine whether the the use of our quality parameters improved the situation recognition rates when we used a transparent reasoning scheme based on voting. We contrasted two approaches to reasoning: : (1) Basic reasoning, where quality was not used, selecting the user activity situation that received the most 'votes' from its causal context events (2) The same voting algorithm but with quality parameters used to attentuate the contribution of each vote.

4.1   Approach

We collected a data set tracking a user in our office. We wanted to determine at any point in time whether a user was 'in' any one of three possible situations: 'busy' at their desk, on a 'break' or at a 'meeting'. To support this, we needed to know where the user was located, whether they were using their computer, and where they were scheduled to be at a meeting. We used three sensors in our dataset: (1) Ubisense location tag sensor that tracks the user's location (2) a computer activity sensor on the user's computer that monitors keyboard and mouse activity and (3) a calendar sensor that indicates whether a meeting is scheduled or not in the user's diary. For example, the situation of user 'busy ' is occurring when the user is at their desk, using their keyboard or mouse and no meeting is scheduled in their diary: 'user hasLocation *desk*', 'computer hasStatus *active*', and 'calendar hasSchedule *free*'. When each of these context events is detected, each contributes a vote towards the 'busy' situation. Situation checking occurs every 30 seconds. The user maintained a diary to track the actual ground truth situations.

When quality parameters are used, each vote is attenuated by the confidence value of its associated context event. In order to determine context event confidence, we needed to determine the underlying sensor and context event quality. We analysed data from the sensors and context events, using our aggregation model in Figure 3. as a basis for finding which quality parameters we should use. Quality parameters for our experiment are shown in Table 1. Values for constant parameters such as reliability are shown. Parameters that are dynamically

Table 1. Quality parameters used

Sensor	Sensor quality	Context event quality
Ubisense	PrecisionX (1.65 m), PrecisionY (1.11m), accuracy (0.8)	Precision membership, reliability (0.72)
Calendar	Precision (10 mins)	Precision membership, reliability (0.6)
Comp. Activity	None used	Fuzzy membership, reliability (0.95)

calculated from sensor readings, such as fuzzy membership, are listed without an accompanying value. The parameters and their values are derived as follows:

- For the Ubisense sensor, precisions (1.65 x-axis, 1.11 y-axis) and accuracy (0.8) were captured using training data where we gathered readings and compared actual location versus sensor readings. We used these precision numbers to generate a precision membership for 'has Location' context events, as described in Section 3.2. An accuracy of 0.8 for Ubisense was degraded at the context event level to a reliability of 0.72 because the user neglected to carry their tag during the data set collection 10% of the time.
- For the calendar sensor, we observed that over a period of a month, the user adhered to 22 out of 36 total meetings in her diary. Therefore, reliability of the calendar sensor is 0.6 based on 60% adherence to diary entries. Start and end times of attended meetings during this time were imprecise by an average of 10 minutes.
- For the computer activity sensor, no noise in sensor readings was observed. However, when abstracted to context event, we applied a fuzzy function to capture the gradual move from active status to inactive and vica versa. The function degraded linearly from fully active (fuzzy membership value of 1) if activity was detected in the last 30 seconds, reducing to 0 after 3 minutes of no activity.
- *Context confidence* for each event was calculated as the product of the context event quality parameter values, as also used by [2]. For example, if at a particular point in time t, a context event of 'user hasLocation desk' has a calculated precision membership of 0.9, and Ubisense reliability is 0.72, the overall confidence of the 'user hasLocation desk' context event at that point in time is 0.65.
- *Situation confidence* for each of the three user situations was calculated as the average of the context event confidences for that situation. For example, at time t, the context events relevant to the user 'busy' situation have the following context event confidences: 'user hasLocation *desk*, conf 0.65', 'calendar hasSchedule *free*, conf 0.9' and 'computer hasStatus *active*, conf 0.4'. Therefore, the confidence of the user 'busy' situation at this point in time t is the average of the three underlying context events, 0.65. At time t, the situation with the highest confidence is deemed to be occurring.

4.2 Analysing Our Results

Situation recognition rates (see Table 4.2) were better when we used our quality model parameters: 90% of situations that could not be identified by the basic reasoning technique were identified when our quality parameters were included in the reasoning process. Basic reasoning failed when the sensor data was noisy (e.g. imprecise Ubisense reading) or user errors were encountered (e.g. user forgot to carry locator tag). The 'meeting' recognition rates were the same because the underlying sensor information was of good quality. I.e. The user always had a meeting scheduled when the meeting was on, the user remembered to wear their Ubisense locator tag to the meeting, and the sensor readings were accurate. Reasoning *with quality parameters* failed for 7% of situation checks for three reasons: (1) The ground truth was captured in a manual diary, with the user rounding up to minutes. This led to mismatches at situation start and end times between ground truth and situation detection. Annotation of ground truth with time in seconds will be important to avoid this problem in future experiments. (2) Some situation rules were inadequate. For example, when a user is reading at their desk without using their computer, the 'busy' at desk situation is not fulfilled because the computer activity sensor is inactive. The inactive computer activity sensor has a higher vote due because its context events are more reliable than the Ubisense related context events, so at 'break' was selected (3) The fuzzy membership of the activity context events led to the system *gradually* recognising that the computer was no longer active. In reality, the user finished the 'busy' situation in an abrupt manner as recorded in the ground truth diary. This requires some thought as to how specific situation transitions and associated application behaviour should work: are instantaneous changes versus gradual changes in situation or behaviour required? Such analysis will help to determine the appropriateness of using fuzzy and decay functions for underlying contexts.

Table 2. Situation recognition results: basic and quality reasoning

Situation	# of readings	basic reasoning: % identified	reasoning with quality: % identified
User busy	720	52	96
User on break	199	75	85
User at meeting	53	91	92

Our results highlighted a number of weaknesses that may be resolved by more a robust reasoning mechanism rather than changes to our quality model. At various points, it was difficult to identify the winning situation because the situation confidences were so close. e.g. How meaningful is it to choose situation of *(busy, conf 0.45)* over *(break, conf 0.44)*? Therefore, an algorithm that can support greater convergence on situation confidence is desirable. It was also clear that conflict between the sensors was manifested by a lack of a clear winner on the voting process. However, the source of conflict is not identified by our voting algorithm. Finally, it was not obvious how to allocate the vote uncertainty. E.g. the contexts events from the calendar sensor had a reliability of 0.6, resulting in 0.4 of the vote being unallocated, which seems intuitively incorrect. We propose that these weaknesses may be addressed by using Dempster Shafer theory as our reasoning mechanism, as discussed in Section 5.

5   Conclusion and Future Work

In this paper, we presented a general context quality model. The structured model in UML provides a flexible and generalised way for designers of context-aware systems to incorporate context quality into their design process; the aggregation model identifies quality issues and their aggregation across context layers. We applied our model to an experimental data set, using a transparent voting algorithm to identify situations. Situation recognition results improved when we used quantified quality measures in our reasoning process.

Currently, we are integrating our quality model work with a more robust uncertainty reasoning algorithm based on Dempster Shafer theory (DST). DST is a mathematical theory of evidence that propagates uncertainty values [12], providing an indication of the quality of inference. We anticipate that use of DST will allow us to address the weaknesses of the voting algorithm identified in 4.2. Early results from this work using a sample of our data set results in situation recognition rates similar to the voting algorithm. However, DST also provides a conflict metric quantifying the extent of context disagreement during situation recognition. Analysis of this conflict may allow us to detect a number of scenarios. We noted that conflict levels were high when sensors are in conflict, rule problems occur or situations are in transition. This is valuable information that we hypothesise can improve our ability to reason with context, particularly in dynamic environments subject to changes in sensors and situations.

References

1. Buchholz, T., Kupper, A., Schiffers, M.: Quality of context: What it is and why we need it. In: Proceedings of the tenth Workshop of the HP OpenView University Association, OVUA 2003 (2003)
2. Dobson, S., Coyle, L., Nixon, P.: Hybridising events and knowledge as a basis for building autonomic systems. IEEE TCAAS Letters (2007)
3. Ebling, M.R., Hunt, G.D.H., Lei, H.: Issues for context services for pervasive computing. In: Proceedings of the Workshop on Middleware for Mobile Computing (2001)
4. Gray, P., Salber, D.: Modelling and using sensed context information in the design of interactive applications. In: Nigay, L., Little, M.R. (eds.) EHCI 2001. LNCS, vol. 2254, pp. 317–335. Springer, Heidelberg (2001)
5. Gu, T., Wang, X.H., Pung, H.K., Zhang, D.Q.: An ontology-based context model in intelligent environments. In: Proceedings of Communication Networks and Distributed Systems Modeling and Simulation Conference, pp. 270–275 (2004)

6. Henricksen, K., Indulska, J.: Modelling and using imperfect context information. In: Proceedings of the 2nd IEEE Annual Conference on Pervasive Computing and Communications Workshops, pp. 33–37 (2004)
7. Korpipaa, P., Mantyjarvi, J., Kela, J., Keranen, H., Malm, E.: Managing context information in mobile devices. IEEE Pervasive Computing 2(3), 42–51 (2003)
8. Lei, H., Sow, D.M., John, I., Davis, S., Banavar, G., Ebling, M.R.: The design and applications of a context service. SIGMOBILE Mob. Comput. Commun. Rev. 6(4), 45–55 (2002)
9. McKeever, S., Ye, J., Coyle, L., Dobson, S.: A multilayered uncertainty model for context aware systems. In: Adjunct proceedings of the international conference on Pervasive Computing: Late Breaking Result, May 2008, pp. 1–4 (2008)
10. Ranganathan, A., Al-Muhtadi, J., Campbell, R.: Reasoning about uncertain contexts in pervasive computing environments. IEEE Pervasive Computing 3(2), 62–70 (2004)
11. Schmidt, A., Aidoo, K.A., Takaluoma, A., Tuomela, U., Laerhoven, K.V., Velde, W.V.D.: Advanced interaction in context. In: Gellersen, H.-W. (ed.) HUC 1999. LNCS, vol. 1707, pp. 89–101. Springer, Heidelberg (1999)
12. Shafer, G.: A mathematical theory of evidence. Princeton Unverisity Press (1976)
13. Wu, H., Siegel, M., Stiefelhagen, R., Yang, J.: Sensor fusion using dempster-shafer theory. In: Proceedings of IEEE Instrumentation and Measurement Technology Conference, Anchorage, AK, USA, May 2002, pp. 7–12 (2002)
14. Ye, J., McKeever, S., Coyle, L., Neely, S., Dobson, S.: Resolving uncertainty in context integration and abstraction: context integration and abstraction. In: ICPS 2008: Proceedings of the 5th international conference on pervasive services, pp. 131–140. ACM, New York (2008)

On a Generic Uncertainty Model for Position Information

Ralph Lange, Harald Weinschrott, Lars Geiger, André Blessing, Frank Dürr,
Kurt Rothermel, and Hinrich Schütze

Collaborative Research Center 627, Universität Stuttgart, Germany
http://www.nexus.uni-stuttgart.de

Abstract. Position information of moving as well as stationary objects
is generally subject to uncertainties due to inherent measuring errors
of positioning technologies, explicit tolerances of position update protocols, and approximations by interpolation algorithms. There exist a
variety of approaches for specifying these uncertainties by mathematical
uncertainty models such as tolerance regions or the Dilution of Precision (DOP) values of GPS. In this paper we propose a principled generic
uncertainty model that integrates the different approaches and derive a
comprehensive query interface for processing spatial queries on uncertain
position information of different sources based on this model. Finally, we
show how to implement our approach with prevalent existing uncertainty
models.

1 Introduction

Position information on moving objects such as mobile devices, vehicles, and
users as well as stationary objects such as buildings, rooms, and furnishings is
an important kind of context information for context-aware applications. The
authors of [1] and [2] even refer to the locations of objects as *primary context*.

Position information is generally subject to uncertainties at every stage of
processing: Already position information acquired by positioning sensors such
as GPS receivers only approximates the actual position of the respective sensor or object due to physical limitations and measurement errors of the sensing
hardware. Update protocols for transmitting position information from sensors
to remote databases or location services further degenerate the position information for the sake of reduced communication cost [3,4,5]. Interpolating in time
between consecutive pairs of position records may result in further uncertainties,
depending on the temporal density of the position information. Fusion of position information on the same phenomenon improves the accuracy but cannot
eliminate uncertainties altogether.

Many context-aware applications must not neglect such uncertainties. For
instance, navigating a blind person around obstacles [6] requires estimates for
the uncertainty of the position information about the blind person as well as
about the obstacles.

K. Rothermel et al. (Eds.): QuaCon 2009, LNCS 5786, pp. 76–87, 2009.

Therefore, a variety of mathematical models for uncertainty of position information have been researched and proposed in the last decades, depending on the specifics and properties of the different technologies and algorithms. For instance, GPS receivers model the distance between sensed and actual position by normal distributions depending on measurement errors and satellite constellation [7]. The authors of [8,9] model all possible positions in-between two given position records by intersecting two circles around the positions given in the records, resulting in a lense-shaped area.

Based on these findings, different *uncertainty-aware* interfaces for accessing and querying position information have been proposed for the different system components such as positioning sensors, update protocol endpoints, moving objects databases, and location services.

With the advent of large-scale context-aware systems such as the Nexus platform [10], applications more and more rely on position information from many different sources. Therefore independent and technology-specific uncertainty models and query interfaces are not sufficient but a generic approach allowing for homogeneous and uncertainty-aware access to position information is required.

Regarding such an approach, we can again distinguish between a generic, mathematically principled model for uncertain position information and a suitable, extended query interface based on this generic uncertainty model. The requirements for the generic uncertainty model are as follows:

- *Expressiveness and generality:* The generic uncertainty model has to be fully compatible with all existing specific uncertainty models for position information of the different positioning sensors, update protocols, and fusion and interpolation algorithms and reflect them with minimal loss of information.
- *Directness:* For simplicity and for minimizing the computational and storage overhead in implementations, the generic uncertainty model has to represent the uncertainty of position information in a self-evident way corresponding to the specific uncertainty models.

The resulting basic requirements for the extended, uncertainty-aware query interface are the following ones:

- *Immediacy and comprehensiveness:* The query interface should immediately build upon the generic uncertainty model to minimize computational effort and exploit all information provided by the uncertainty model.
- *Generality:* The query interface has to provide all prevalent spatial query types for position information such as position query, range query, and next-neighbor query, cf. [11].

In this paper we propose a generic uncertainty model based on partial spatial distribution functions and a corresponding extended query interface supporting five prevalent query types satisfying the above requirements. Our approach is suitable for all (uncertain) point-shaped position information in the two-dimensional space.

In detail, we present the following contributions: In Section 2, we survey existing, specific uncertainty models for position information. In Section 3, we show that they can be classified into three fundamental types and that they all base on partial spatial distribution functions and then derive our generic uncertainty model. Based on this finding we present an extended query interface for uncertain position information in Section 4 and show how to implement this interface for different specific uncertainty models. Section 5 discusses related work, before we present our conclusions and outlook in Section 6.

2   Survey of Specific Uncertainty Models

In this section, we survey the existing uncertainty models for position information. Most of these models only consider two-dimensional positions. Height information of indoor positioning systems is often reduced to information about the floor and, in case of outdoor systems such as GPS, the height information is handled separately. Therefore, we restrict this survey to two-dimensional positions.

2.1   Uncertainty Models of Positioning Systems

Positioning systems use different techniques like triangulation/-lateration, scene analysis, or proximity sensing to determine positions [12]. Different positioning techniques not only yield different scales of accuracy—from millimeters to hundreds of meters—but also result in different uncertainty models.

Positioning systems based on trilateration mostly model the position sensed at time t and denoted by s_t as a normal distribution. This particularly applies to global navigation satellite systems such as GPS [7] and ultrasonic-based positioning systems such as Cricket [13]. For instance, according to [7] the standard deviation σ of a two-dimensional position determined by GPS can be calculated based on the User Equivalent Range Error σ_{UERE} and the Horizontal Dilution of Precision, HDOP, as

$$\sigma = \text{HDOP} \cdot \sigma_{\text{UERE}} \tag{1}$$

Other systems—e.g., the WiFi positioning system presented in [14]—only give a center point and several percentile values around that point, i.e. s_t consists of several concentric circles expressing the probability that the actual position, denoted by a_t, lies within a given circle. In beaconing systems, such as Active Badge [15], the sensed position may only specify an area such as a room—i.e. a_t is known to be in that area but without any further distribution information. The same applies to the Smart Floor positioning systems using pressure-sensitive tiles [16] and positioning using passive RFID tags [17].

2.2   Modeling of Uncertainty in Update Protocols

Update protocols introduce further uncertainties to position information and lead to new uncertainty models. Dead reckoning protocols trade accuracy off against communication cost for efficient transmission of position information from a remote positioning sensor to stationary components managing the current position [3,4]. Remote trajectory simplification algorithms additionally consider the costs for storing the current and past positions, i.e. the whole trajectory [18,5]. For all these approaches, a position s_t is modeled by a center point and a distance value for the maximum distance from that point. However, the distribution within the resulting circle is undefined.

2.3   Uncertainty Models for Fusion and Interpolation

Fusion algorithms improve the accuracy of a sensed position from different sensor data on the same phenomenon. Multi-Area Probability-based Positioning by Predicates [19] describes s_t by a number of polygons with probability values taking into account even multiple predicates on the position. For fusion of arbitrary probability density functions different Bayes filter implementations are applicable, possibly discretizing the plane using a grid [20].

Complex, uncertainty-aware interpolation algorithms allow for deriving positions at times in-between two sensing operations taking into account the temporal discretization introduced by sensing. The authors of [8,9] show how to restrict the position s_t at such a time to a lense-shaped area—the intersection of two circles—by means of the maximum speed between the sensing operations.

3   Mathematical Generalization of Uncertainty Models

The diversity of these specific models makes it difficult to incorporate uncertain position information from different sources in applications. We show in the next section, however, that all these different models can be reduced to a common mathematical model for uncertain position information. After that, we propose a consistent interface for applications that need to access uncertain position data from different specific models. This interface is based on our common mathematical model.

3.1   Classification of Uncertainty Models

Although a large number of different specific models for uncertain position information exists, we can classify them into three major types as illustrated in Figure 1:

1. *pdf-based models:* These models use complete probability density functions to describe the uncertainties of positions. Hence, with such a model, a position at time t is described by a two-dimensional probability density function $s_t : \mathbb{R}^2 \rightarrow [0, \infty)$.

    Amongst others, pdf-based models are used for specifying the uncertainty of trilateration-based positioning systems such as GPS.

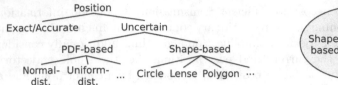

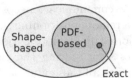

Fig. 1. Taxonomy of major classes of the existing uncertainty models

Fig. 2. Mathematical generality of the major uncertainty models

2. *Shape-based models:* Models of this class describe positions by geometric shapes. These shapes have associated probability values, however, in contrast to the pdf-based models, the approaches make no claims about the probability distribution within a shape.

   Hence, a position at time t is a set $s_t = \{(A_1, p_1), \ldots, (A_n, p_n)\}$ where $p_j \in [0,1]$ and $A_j \subseteq \mathbb{R}^2$ are geometric shapes such as polygons or circles.

   Shape-based models are used, for example, for position information from infrared beacons, RFID tags, or interpolation with the intersection of circles.

3. *Accurate model:* For completeness, we also include the accurate model for specifying an exact position without uncertainty.

   Formally, a position is described by a single point, representing the actual position, i.e. $s_t = a_t$.

3.2   Generic Uncertainty Model

In terms of probability theory, all three classes of uncertainty models provide probabilistic information on the actual position they describe as they allow for mapping from one or more geometric shapes A to cumulative probabilities p. More precisely, they describe the position at time t by a (generally partial) function $s_t : \mathcal{P}(\mathbb{R}^2) \to [0,1]$ with

$$s_t(A) = P[a_t \in A] = p \,. \tag{2}$$

We refer to such a function as partial spatial distribution function (psdf). Note that pdf-based models even allow for computing a mapping for all $A \in \mathcal{P}(\mathbb{R}^2)$ and thus can be treated as special, non-partial cases of psdf. Accurate positions given by the accurate model likewise are special cases of psdf, where $s_t(A) = 1$ if $a_t \in A$, and otherwise $s_t(A) = 0$.

Thus, regarding psdf, the three classes of uncertainty models can be nested according to their mathematical generality as illustrated in Figure 2.

It holds that $A_j \subseteq A_k$ implies $s_t(A_j) \leq s_t(A_k)$ as well as $s_t(\mathbb{R}^2) = 1$. Thus, given an arbitrary area A, a psdf s_t allows for deriving two estimates p_{lower} and p_{upper} with $0 \leq p_{\text{lower}} \leq s_t(A) \leq p_{\text{upper}} \leq 1$ for the position of the corresponding object at time t.

The three major classes of uncertainty models and their common basis in terms of probability theory is an important finding and composes the generic uncertainty model for the extended query interface proposed in the next section.

As the generic uncertainty model includes all classes of specific uncertainty models, it satisfies the requirements *expressiveness* and *generality* given in Section 1. Furthermore, it satisfies *directness* as it immediately bases on shape-based models, the most general class of uncertainty models in mathematical terms.

4   Uncertainty-Aware Query Interface

In this section we present an extended, uncertainty-aware query interface for position information based on the above finding of a generic uncertainty model. Therefore, the query interface can be implemented for every existing uncertainty model and thus allows for uncertainty-aware processing of position information from different, heterogeneous sources.

Next, we discuss the extended, uncertainty-aware versions of prevalent spatial query types for position information and thus show that the query interface meets the requirement *generality* given in Section 1. Then, we describe a number of examples how to implement the query types and thereby the query interface for different specific uncertainty models, which shows that the query interface also satisfies the requirements *immediacy* and *comprehensiveness*.

4.1   Extended Query Types

In the following, we consider an arbitrary set of moving or stationary objects $\{O_1, \ldots, O_n\}$. Though most entities of the real-world have a certain extent, we only consider point objects. For any given object, one can always define an anchor point and thus reduce its position to this point. We denote the position of object O_i at time t by $s_{i,t}$. All queries have two parameters O_i and t in common, specifying the queried object and the queried time, respectively.

Besides the uncertain position information, we argue that the providers must define most likely point positions for each time and object they manage. This *defined point* $c_{i,t}$ for an object O_i at time t may be either modeled explicitly or computed on the fly from the uncertain position information of O_i. Note that this point is naturally given with most existing uncertainty models such as normal distributions or circular shapes.

The defined point $c_{i,t}$ of O_i at time t serves to define an unambiguous mapping $\bar{c}_{i,t} : [0,1] \to \mathcal{P}(\mathbb{R}^2)$ from each cumulative probability p to the circular area $\bar{c}_{i,t}(p) = A^C$ with center $c_{i,t}$ and minimum radius such that $s_{i,t}(A^C) \geq p$. This is needed, as the inverse $s_{i,t}^{-1}(p)$ is generally ambiguous. For instance consider the 2D normal distribution illustrated in Figure 4: The left half, the right half, and the inner circle are three examples of areas with $s_{i,t}(A) = 0.5$.

Where applicable, the circular areas $\bar{c}_{i,t}(p)$ are clipped to $A_{i,t}^1$, the smallest area with $s_{i,t}(A_{i,t}^1) = 1$, which always is unambiguous but may be equal to $\mathbb{R}^2$.

Position Query. Besides O_i and t, the position query takes a parameter $p \in [0,1]$ and returns the smallest area $A = \bar{c}_{i,t}(p) \cap A_{i,t}^1$ such that $s_{i,t}(A) \geq p$:

$$\text{Position Query:}\quad \text{PQ}\,(O_i, t, p) \to (A, \bar{c}_{i,t}(p), s_{i,t}(A)) \tag{3}$$

Moreover, it returns $\bar{c}_{i,t}(p)$ and the probability value $s_{i,t}(A)$.

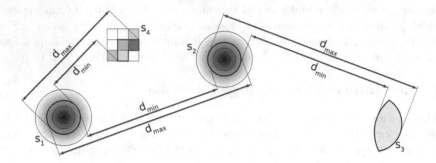

Fig. 3. Distance query evaluation

Inside and Range Query. To test whether an object is within an area A with a probability of at least $p_{\text{true}} > p_{\text{false}}$, the inside query is defined as:

$$\text{Inside Query:} \quad \text{IQ}(O_i, t, A, p_{\text{true}}, p_{\text{false}}) \rightarrow (\{\texttt{true}, \texttt{maybe}, \texttt{false}\}) \quad (4)$$

With the estimates for p_{lower} and p_{upper} from Section 3, the inside query returns **true** iff $p_{\text{lower}} \geq p_{\text{true}}$ and **false** iff $p_{\text{upper}} \leq p_{\text{false}}$. In all other cases, the uncertain position obviously overlaps the area A as well as its inverse $\mathbb{R}^2 \setminus A$ and the query returns **maybe**.

The range query can be implemented easily by inside queries on the set of queried objects.

Distance and Nearest-Neighbor Query. The distance query returns an upper and lower bound for the distance between two objects with a minimum probability of p by computing the minimum and maximum distances $d_{\min}$ and $d_{\max}$ between the two shapes $\bar{c}_{i,t}(p) \cap A^1_{i,t}$ and $\bar{c}_{j,t}(p) \cap A^1_{j,t}$.

$$\text{Distance Query:} \quad \text{DQ}(O_i, O_j, t, p) \rightarrow (d_{\min}, d_{\max}) \quad (5)$$

Figure 3 illustrates several examples of how $d_{\min}$ and $d_{\max}$ are computed. In Section 4.2, these examples are discussed in detail.

The nearest-neighbor query uses these distance bounds to derive the set of objects that may be closest to a given object O_i. Given O_i and a probability value p, it computes the pairwise distances between O_i and all other objects O_j ($i \neq j$) as described above and determines the maximum distance $\hat{d}$ for O_i's nearest neighbor as $\hat{d} = \min(d_{\max})$ for all $O_j \neq O_i$. Then, it returns the set of objects with their distance bounds that may be closer to O_i than $\hat{d}$:

$$\text{Nearest-Neighbor Query:} \quad \text{NNQ}(O_i, t, p) \rightarrow (O_j, d_{\min}, d_{\max})^* \text{ where } d_{\min} \leq \hat{d}$$

Thus, depending on p, the result either contains only objects that are likely to be nearest neighbors or also objects with a low probability of being nearest neighbor.

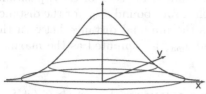

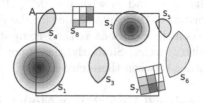

Fig. 4. 2D normal distribution   **Fig. 5.** Range query evaluation

4.2   Implementing the Query Interface

In the following, we exemplary discuss how to implement the proposed query interface for three specific uncertainty models. For actual implementations different spatial data models such as the simple feature types of the Open Geospatial Consortium (OGC) are feasible[1].

As a first example, we consider the uncertainty model of GPS [7] based on normal distributions. Then, we discuss the lense-based uncertainty model of the interpolation algorithm in [8,9]. Finally, we show how to map a grid-based uncertainty model [20] to the extended query interface. For all of these models, we show how to implement PQ, IQ, and DQ. We leave out RQ and NNQ since these are straight-forward extensions of IQ and DQ.

GPS Uncertainty Model Based on Normal Distribution. A GPS position is given by longitude, latitude, and the HDOP value specifying a 2D normal distribution. Longitude and latitude can be directly used as defined point $c_{i,t}$ for the generic model. As already described in Equation 1, the HDOP value is multiplied with the device-specific User Equivalent Range Error, σ_{UERE}, to derive the standard deviation σ of the normal distribution [7].

As an example consider a GPS sensor with accuracy $\sigma_{\text{UERE}} = 5\,\text{m}$ reporting $(lat, lon, HDOP)$ as $(48°47'N, 9°11'O, 1.5)$. In this case, a PQ with $p = 0.75$ returns a circle A^C that is centered at $(48°47'N, 9°11'O)$ with radius $r = 8.84\,\text{m}$ by solving the circular integral over the density function $f_{N^2(0,\sigma^2)}(x,y)$ of the two-dimensional normal distribution with $\sigma = \text{HDOP} \cdot \sigma_{\text{UERE}} = 1.5 \cdot 5\,\text{m} = 7.5\,\text{m}$ for r, i.e. by solving

$$p = \int_{\sqrt{x^2+y^2} \leq r} f_{N^2(0,\sigma^2)}(x,y)\,\mathrm{d}(x,y) \ . \tag{6}$$

Figure 5 shows a queried range A and the positions of eight objects. s_1 and s_2 are GPS positions where an IQ can be unambigously evaluated by integrating $f_{N^2(0,\sigma^2)}(x,y)$ over A. For $p_{\text{true}} = 0.8$ and $p_{\text{false}} = 0.2$, IQ returns maybe for s_1 and true for s_2.

Figure 3 shows several positions of objects and the upper and lower bound for the distances between pairs of these positions. In case of a DQ on two GPS

---

[1] Depending on the data model, curves (e.g., of circles or lenses) have to be approximated by polygons at a suitable level of granularity.

positions s_1 and s_2 with $p = 0.75$, A^C is computed for each of these positions according to the explanations for the PQ. The lower bound d_{min} for the distance between the positions is then computed as the minimal distance between the resulting circles. Similarly, the upper bound d_{max} is computed as the maximal distance between these circles.

Lense-based Uncertainty Model. For interpolation with lenses [8,9] (cf. Section 2), we consider two consecutive position fixes of an object O_i at times $t_1 = 0\,\mathrm{s}$ and $t_2 = 100\,\mathrm{s}$ in the Euclidean plane with $s_{i,t_1} = (0\,\mathrm{m}, 0\,\mathrm{m})$ and $s_{i,t_2} = (100\,\mathrm{m}, 0\,\mathrm{m})$. In addition, we assume the maximum speed of the object is known to be $1.5\,\mathrm{m/s}$. The position query PQ for time $t = 50\,\mathrm{s}$ returns the intersection of the circles centered at s_{i,t_1} and s_{i,t_2} with radius $r = 1.5\,\mathrm{m/s}{\cdot}50\,\mathrm{s} = 75\,\mathrm{m}$ for any queried p. Note that the probability $s_{i,t}(A)$ given in the query result always is 1.0.

Also note that any point within the lense can be chosen as defined point $c_{i,t}$ without affecting the result $A = \bar{c}_{i,t}(p) \cap A^1_{i,t}$ of the PQ (cf. Equation 3). An obvious choice for $c_{i,t}$ is the linear interpolation between s_{i,t_1} and s_{i,t_2}.

For the queried range A given in Figure 5, the IQ returns true for position information s_3, false for s_6, and maybe for s_4 and s_5 for any value of p specified in the queries.

Figure 3 shows an example for a DQ involving a lense-based position s_3. To evaluate the DQ between s_3 and the GPS position s_2 with $p = 0.75$, a PQ on s_3 is processed, which results in a lense shape. Then, the lower bound d_{min} for the distance between the positions is computed as the minimal distance between the lense shape of s_3 and A^C of s_2. Similarly, the upper bound d_{max} is computed as the maximal distance.

Grid-based Uncertainty Model. Consider a grid-based [20] position that is given by a set of tuples (x_j, y_j, p_j) where x_j, y_j are cell coordinates and p_j is the corresponding probability. Thus, the grid-based position is given by a set of disjoint quadratic shapes with associated probabilities. As defined point $c_{i,t}$ for the generic uncertainty model, a couple of alternatives are conceivable: First, the center of the cell with highest probability p_j is chosen. Second, the defined point is selected as the centroid.
As an example consider a grid-based position defined by

$$s_{i,t} = (1,1,0.15), (2,1,0.05), (2,2,0.2), (3,2,0.5), (3,3,0.1) .$$

We compute the centroid (x_c, y_c) of this position as defined point by taking the weighted sum of cell indices over each dimension:

$$x_c = 1 \cdot 0.15 + 2 \cdot (0.05 + 0.2) + 3 \cdot (0.5 + 0.1) = 2.45$$
$$y_c = 1 \cdot (0.15 + 0.05) + 2 \cdot (0.2 + 0.5) + 3 \cdot 0.1 = 1.9$$

A PQ with $p = 0.75$ is evaluated by selecting the cells closest to the centroid until the aggregated probability of the cell equals or exceeds 0.75. In this example the polygon enclosing the cells $(2, 1, 0.05), (2, 2, 0.2), (3, 2, 0.5)$ is returned. For $s_{i,t}(A)$, a value of 0.75 is returned, since the sum of these cells' probabilities equals 0.75.

For the queried range A in Figure 5, the IQ for the grid-based position information s_7 can be evaluated unambiguously since the range A is aligned to its grid. As the grid of position s_8 is not aligned to the range A, p_{upper} and p_{lower} differ. p_{lower} is the sum of probabilities of the cells that are covered by range A. In contrast, p_{upper} is evaluated as the sum of probabilities of cells that overlap with A.

Figure 3 shows an example for a DQ involving a grid-based position s_4. The processing of the DQ between s_4 and the GPS position s_1 with $p = 0.75$ is based on the result of a PQ on s_4. With the grid-based uncertainty model, a PQ results in a polygonal area possibly consisting of multiple unconnected parts. The lower bound $d_{\min}$ for the distance is computed as the minimal distance between A^C of s_1 and the nearest part of the area returned by the PQ. The upper bound $d_{\max}$ is computed as the maximal distance to the most distant part.

5 Related Work

The proposed query interface for uncertain position information and its generic uncertainty model relates to two fields: Models for uncertain spatial data in general and specific approaches for uncertain position information of moving objects.

Pauly and Schneider [21] classify the former into models based on rough sets like the Egg-Yolk approach [22] and models based on fuzzy sets like the fuzzy spatial data types proposed in [23]. The models particularly define topological predicates for vague spatial regions but do not aim at a generic model integrating the variety of existing uncertainty models.

A variety of algorithms for processing range and next-neighbor queries on uncertainty position information have been proposed in recent years, e.g., [9,24,25]. Most approaches model uncertain positions as circular areas which can be mapped to the proposed generic model. Moreover, they use compatible semantics for query results such as the MAY/MUST semantics for the containment in queried regions proposed in [25].

Existing approaches for fusion of position sensor data—particularly Bayesian filtering [20] and inferring from location predicates [19]—are also covered by the generic uncertainty model as discussed in the previous sections.

6 Conclusions and Outlook

In this paper we discussed the need for a generic uncertainty model for position information in large-scale context-aware systems and formulated the requirements for a suitable model and uncertainty-aware query interface.

We surveyed and classified the variety of existing technology-specific uncertainty models and showed that they all can be considered as partial spatial distribution functions (psdf) with respect to their mathematical generality.

Based on this finding, we proposed an extended query interface for position information by extending common query types with information on the position uncertainty. Furthermore, we discussed how to implement these types for certain prevalent uncertainty models. These examples show that the proposed query interface can provide homogeneous access to uncertain position information from different sources and sensors and that the proposed approach meets the various requirements formulated in Section 1.

Although we only discussed position information in this paper, the mathematical approach can be extended easily to scalar data types (e.g., temperature and velocity) as well as data with three or more dimensions. Of course, the query interface has to be adapted to the relevant query types for these data types.

Currently, we only consider firm boundaries for the queried ranges with inside and range queries. In the future, we will expand our approach to also support ranges with uncertainties.

Acknowledgments. The work described in this paper was partially supported by the German Research Foundation (DFG) within the Collaborative Research Center (SFB) 627.

References

1. Schmidt, A., Beigl, M., Gellersen, H.W.: There is more to context than location. Computers & Graphics Journal 23, 893–902 (1999)
2. Dey, A.K., Abowd, G.D.: Towards a better understanding of context and context-awareness. In: Proc. of CHI 2000 Workshop on the What, Who, Where, When and How of Context-Awareness, The Hague, Netherlands (2000)
3. Leonhardi, A., Rothermel, K.: A comparison of protocols for updating location information. Cluster Computing: The Journal of Networks, Software Tools and Applications 4, 355–367 (2001)
4. Čivilis, A., Jensen, C.S., Pakalnis, S.: Techniques for efficient road-network-based tracking of moving objects. IEEE Trans. on Know. and Data Eng. 17, 698–712 (2005)
5. Lange, R., Farrell, T., Dürr, F., Rothermel, K.: Remote real-time trajectory simplification. In: Proc. of 7th PerCom, Galveston, TX, USA (2009)
6. Hub, A., Kombrink, S., Bosse, K., Ertl, T.: Tania – a tactile-acoustical navigation and information assistant for the 2007 csun conference. In: Proc. of 22nd CSUN, Los Angeles, CA, USA (2007)
7. United States Department of Defense, Navstar GPS: Global Positioning System Standard Positioning Service Performance Standard, 4th edn (2008)
8. Pfoser, D., Jensen, C.S.: Capturing the uncertainty of moving-object representations. In: Proc. of 6th SSD, Hong Kong, China, pp. 111–131 (1999)
9. Cheng, R., Kalashnikov, D.V., Prabhakar, S.: Querying imprecise data in moving object environments. IEEE Trans. on Know. and Data Eng. 16, 1112–1127 (2004)
10. Lange, R., Cipriani, N., Geiger, L., Großmann, M., Weinschrott, H., Brodt, A., Wieland, M., Rizou, S., Rothermel, K.: Making the world wide space happen: New challenges for the nexus context platform. In: Proc. of 7th PerCom, Galveston, TX, USA, pp. 300–303 (2009)

11. Güting, R.H., Schneider, M.: Moving Objects Databases. Morgan Kaufmann Publishers, San Francisco (2005)
12. Hightower, J., Borriello, G.: Location systems for ubiquitous computing. Computer 34, 57–66 (2001)
13. Smith, A., Balakrishnan, H., Goraczko, M., Priyantha, N.: Tracking moving devices with the cricket location system. In: Proc. of 2nd MobiSys, pp. 190–202 (2004)
14. Bahl, P., Padmanabhan, V.N.: Radar: An in-building rf-based user location and tracking system. In: Proc. of 19th INFOCOM, pp. 775–784 (2000)
15. Harter, A., Hopper, A.: A distributed location system for the active office. IEEE Network 8, 62–70 (1994)
16. Orr, R.J., Abowd, G.D.: The smart floor: A mechanism for natural user identification and tracking. In: Extended Abstracts on Human factors in Computing Systems (CHI 2000), The Hague, The Netherlands, pp. 275–276 (2000)
17. Mehmood, M.A., Kulik, L., Tanin, E.: Autonomous navigation of mobile agents using rfid-enabled space partitions. In: Proc. of 16th ACM GIS, Irvine, CA, USA (2008)
18. Lange, R., Dürr, F., Rothermel, K.: Online trajectory data reduction using connection-preserving dead reckoning. In: Proc. of 5th MobiQuitous, Dublin, Ireland (2008)
19. Roth, J.: Inferring position knowledge from location predicates. In: Hightower, J., Schiele, B., Strang, T. (eds.) LoCA 2007. LNCS, vol. 4718, pp. 245–262. Springer, Heidelberg (2007)
20. Fox, D., Hightower, J., Liao, L., Schulz, D., Borriello, G.: Bayesian filtering for location estimation. IEEE Per. Comp. 2, 24–33 (2003)
21. Pauly, A., Schneider, M.: Topological predicates between vague spatial objects. In: Bauzer Medeiros, C., Egenhofer, M.J., Bertino, E. (eds.) SSTD 2005. LNCS, vol. 3633, pp. 418–432. Springer, Heidelberg (2005)
22. Cohn, A.G., Gotts, N.M.: The 'egg-yolk' representation of regions with indeterminate boundaries. In: Proceedings GISDATA Specialist Meeting on Spatial Objects with Undetermined Boundaries, pp. 171–187 (1996)
23. Schneider, M.: Fuzzy Spatial Data Types for Spatial Uncertainty Management in Databases. In: Handbook of Research on Fuzzy Information Processing in Databases, pp. 490–515 (2008)
24. Trajcevski, G., Wolfson, O., Zhang, F., Chamberlain, S.: The geometry of uncertainty in moving objects databases. In: Jensen, C.S., Jeffery, K., Pokorný, J., Šaltenis, S., Bertino, E., Böhm, K., Jarke, M. (eds.) EDBT 2002. LNCS, vol. 2287, pp. 233–250. Springer, Heidelberg (2002)
25. Yu, X., Mehrotra, S.: Capturing uncertainty in spatial queries over imprecise data. In: Mařck, V., Retschitzegger, W., Štěpánková, O. (eds.) DEXA 2003. LNCS, vol. 2736, pp. 192–201. Springer, Heidelberg (2003)

A Probabilistic Filter Protocol
for Continuous Queries

Jinchuan Chen, Reynold Cheng, Yinuo Zhang, and Jian Jin

Department of Computer Science, The University of Hong Kong
Pokfulam Road, Hong Kong
{jcchen,ckcheng,ynzhang,jjin}@cs.hku.hk
http://www.cs.hku.hk/

Abstract. Pervasive applications, such as location-based services and natural habitat monitoring, have attracted plenty of research interest. These applications make use of a large number of remote positioning devices like Global Positioning System (GPS) for collecting users' physical locations. Generally, these devices have battery power limitation. They also cannot report very accurate position values. In this paper, we consider the evaluation of a long-standing (or continuous) query over inaccurate location data collected from positioning devices. Our goal is to develop an energy-efficient protocol, which provides some degree of confidence on the query answers evaluated on imperfect data. In particular, we propose the *probabilistic filter*, which governs GPS devices to decide upon whether location values collected should be reported to the server. We further discuss how these filters can be developed. This scheme reduces the cost of transmitting location updates, and hence the energy spent by the GPS devices. It also allows some portion of query processing to be deployed to the devices, thereby alleviating the processing burden of the server.

Keywords: location uncertainty, continuous probabilistic queries, probabilistic filter.

1 Introduction

Recent advances in positioning technologies (e.g., GPS installed in mobile devices and cellular systems) and wireless networks have led to a tremendous growth of location-based applications. For example, a transportation monitoring system may collect the location data provided by vehicles' GPS devices, in order to perform online monitoring or traffic pattern analysis on some geographical area of interest. Interesting applications, such as asking "is there any friend close to my location?" can also be developed.

An important problem faced by these applications is that the location data are inherently uncertain, due to factors like inaccurate measurements, limited sampling rates, and network latency. Services that base their decisions on these data must take this error information into account or else risk the degradation

K. Rothermel et al. (Eds.): QuaCon 2009, LNCS 5786, pp. 88–97, 2009.

of their quality and reliability. Another issue is that generating and transmitting location data requires energy consumption. In mobile devices, where battery power is a scarce resource, these data reporting activities should be carefully controlled, in order to enable these devices to have a long lifetime.

In such environments, it is common to execute long-standing, or *continuous* queries, on the positions of moving vehicles or people. An example of continuous queries is: "Report to me the locations of vehicles in a designated area within the next hour". Such queries allow users to perform real-time tracking of locations of moving objects. To enable the successful execution of these queries, the problems of data uncertainty and energy usage that are prevalent in location-based applications need to be treated with care. In this paper, we study a protocol that enables continuous queries to be *accurately* answered, as well as *efficient* in terms of energy use.

Specifically, to handle location uncertainty of a mobile object, we adopt the *attribute uncertainty* model [13], which assumes that the actual data value is located within a closed region, called the *uncertainty region*. In this region, a non-zero probability density function (pdf) of the value is defined, such that the integration of pdf inside the region is equal to one. Figure 1 illustrates that the uncertain value of a mobile object's location follows a normalized Gaussian distribution [13,16]. Based on this model, we propose the *continuous probabilistic queries* (or CPQ in short), which evaluate attribute uncertainty [1]. These queries differ from their traditional counterparts in that they provide *probabilistic* guarantees for the answers returned. An example of a CPQ can be one that requests the system to report the names and corresponding probabilities of soldiers who are located in the enemy base in the coming 24 hours. One such answer at a particular time instant can be: $\{(o_a, 0.8), (o_b, 0.1)\}$, where o_a and o_b are the IDs of two soldiers. Although we do not know exactly whether o_a and o_b are in the enemy base, we have a relatively higher confidence about o_a's presence in that area than that of o_b. These query answers are probabilistic due to the existence of location uncertainty.

Although CPQ provides a certain level of answer correctness (compared to the case when uncertainty is not considered at all), it is expensive to evaluate. Notice that although a mobile object can report its location value periodically to the system, a lot of energy resources are drained from the object. Moreover, when a location value is received, the system has to recompute the answers to a CPQ. Using our previous example, suppose that o_a moves away from the enemy base,

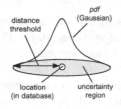

Fig. 1. Uncertainty of location

while o_b, and another object o_c, move towards it. The new query answer becomes: $\{(o_a, 0.3), (o_b, 0.6), (o_c, 0.1)\}$. Recomputing these probability values typically require costly numerical integration [2], and place severe burden to the system. An important problem is then how to evaluate CPQ in an efficient manner.

Our goal is to develop a protocol that can reduce the effort of evaluating CPQs. The main idea is to use the concept of *probabilistic filters*. Generally, a filter is a set of conditions deployed to a mobile object, which governs when the mobile object should report its location value to the system, without violating query correctness [12,3]. Instead of periodically reporting its location values, an object *only* does so only if this is required by the filter installed on it. Hence, the amount of data sent by the mobile object, as well as the energy spent, can be significantly reduced. We will discuss how these filters can be defined for mobile objects, for a particular type of CPQ.

The rest of this paper are organized as follows. Section 2 presents the related work.We decribe the problem setting and the query to be studied in Section 3. Then we discuss the filter protocol in Section 4. Section 5 concludes the paper.

2   Related Works

Uncertainty Management. Recently, a number of uncertain data models have been proposed [16,1,5]. Here we use the attribute-uncertainty model, which generally represents the inexactness in the attribute value as a range of possible values and a pdf bounded in the range [16,1]. Based on this model, the notion of probabilistic queries has been proposed. These are essentially spatial queries that evaluate uncertain location data, and augment probabilistic guarantees to the query result [18,13,1]. There has been plenty of active research on probabilistic queries. [1] proposed a classification scheme of probabilistic queries based on whether a query returns a numerical value or the identities of the objects that satisfy the query. Since computing probabilities can involve expensive integration operations on the pdfs, many studies focus on reducing the computation time of probabilistic queries [4,2,11].

Most of the work on probabilistic queries focuses on snapshot queries - queries that are only evaluated by the system once. Few studies have considered the execution of continuous probabilistic queries (CPQ) on uncertain data. In [8], the selection of an appropriate number of sensors to yield a high-quality result for a CPQ is studied. We study the important problem of handling the communication and computational burden involved in executing a CPQ.

Continuous Query Processing. There has been rapid growth in the development of location management systems [10,6]. The main goal of such systems is to provide efficient location-based services, including road-traffic monitoring and mobile commerce.

The handling of frequent data updates and continuous queries places heavy load on the system. Thus, a number of approaches have been proposed to address these problems, including efficient indexing schemes that can be adapted to the high frequency of data updates [14], incremental algorithms for reducing the query reevaluation cost [19], and the use of adaptive safe regions for reducing update costs [9].

To further reduce the system load, some researchers have proposed to deploy some processing to remote streaming sources capable of performing computation. Here the idea of stream filters can be used. Specifically, each data object is installed with some simple conditions, e.g. filter constraints, that are derived from requirements of a continuous query [12,3,15,7]. The remote object sends its data to the central server only if this value violates the filter constraints. This means that not all values are sent to the server, and a lot of communication effort can be saved.

Note that most of the processing techniques for continuous queries assume that data values are precise. In this paper we consider filtering techniques for uncertain location data.

3   Problem Definition

We now present the formal problem setting. We describe the semantics of the query to be studied, and a naive solution for evaluating such a query.

Framework. As shown in Figure 2, the system consists of a server, which allows users to register their query requests. We study *continuous probabilistic queries* (or CPQ in short), which are long-standing queries that handle location uncertainty. The *query manager* is responsible for evaluating CPQs. Moreover, it makes sure that the query results at the users' site are up-to-date. The result of a continuous query can change due to the update of the mobile objects being tracked.

The data evaluated by the query manager comes from the *uncertain database*, which records attribute uncertainty of data sent by mobile objects (Figure 1). The uncertain database is also used to fetch the location data reported from the mobile objects. Each object collects its location value and the associated uncertainty information. The server also contains a *filter manager*, whose primary

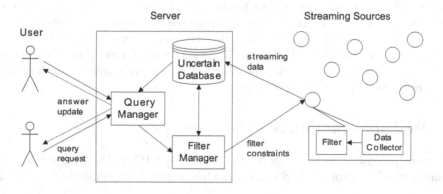

Fig. 2. System Architecture

purposes are to reduce the network bandwidth and energy utilization of mobile objects. This is done by deploying filter constraints to the objects. As will be detailed in Section 4, these filter constraints are computed by the server based on the query information and uncertainty of each data object.

Let $o_1, \ldots, o_n$ be the n mobile objects monitored by the system. The semantics of CRQ studied here are defined as follows:

Definition 1. Continuous Probabilistic Range Query (CPRQ): *Given a time interval* $[t_1, t_2]$, *and a real value* $P \in [0, 1]$, *a CPRQ returns a set of objects* $\{o_i | p_i(t) \geq P\}$ *at every time instant* t, *where* $t \in [t_1, t_2]$, *and* $p_i(t)$ *is the probability that the location of* o_i *is inside a closed region* R.

Here, p_i is called the *qualification probability*, and P is the *probability threshold* [4,17]. An example of CPRQ is: "During the period between $1PM$ and $2PM$, what are the IDs of vehicles that are inside district R, with qualification probability higher than $P = 30\%$?". The CPRQ then continuously report and update a list of vehicle IDs to the query user.

To answer CPRQ at time t, a naive method is to compute the qualification probability of each mobile object o_i, using the following formula:

$$p_i(t) = \int_{u_i(t) \cap R} f_i(x, t) dx \tag{1}$$

In Equation 1, $u_i(t)$ is the uncertainty region of o_i at time t (see Figure 1), and $u_i(t) \cap R$ denote the overlapping part of u_i and the query region R. Also, x is a two-dimensional vector that denotes a possible location of o_i, and $f_i(x, t)$ is the probability density function (pdf) of x. At each time instant t, if an object's location is generated by the object, then its new location is immediately sent to the server, and its probability is computed using Equation 1. After all $p_i(t)$'s have been computed at time t, the IDs of objects whose qualification probabilities are not smaller than P are returned to the user. Using this method, the query answer is constantly refreshed within time interval t_1 and t_2.

Although the above method can answer CPRQ correctly, it suffers from two major problems. First, every object has to report its location once it is generated. This requires a large amount of energy resources. Secondly, the server has to refresh the answers and send them to the end user, whenever it receives a new update from the user. Sometimes, this is not necessary, for example, if the object under consideration still satisfies the query. Next, we study how the probabilistic filter protocol can handle these issues.

4   Probabilistic Filters

We now present a better solution for answering CPRQ. We first outline the framework of the protocol in Section 4.1. Then, we study its details in Section 4.2.

4.1 Protocol Framework

As shown in Figure 2, the server contains a *filter manager*. Its purpose is to control the amount of data updates from the mobile objects, while ensuring the query correctness requirements are satisfied. Specifically, Algorithm 1 shows how the filter manager of the server interacts with the objects, after a query, called q, is submitted to the system. It also shows the algorithm running behind each mobile object o_i:

Algorithm for Processing a CPRQ

```
procedure Server-Side(query q)
1.      Collect current locations from all objects;
2.      Compute new filter constraint for q;
3.      for every object o_i
4.         Send(add_filter_constraint, o_i);
5.      while (t_1 <= current_time <= t_2)
6.         wait for update from object o_i;
7.         Update_uncertain_DB(o_i);
8.         if (update == (o_i, l_i, delete)) // l_i is location of o_i
9.            remove o_i from query answer;
10.        else
11.           insert o_i to query answer;
12.     for every object o_i  // q has finished execution
13.        Send(delete_filter_constraint, o_i);

procedure Object-Side(object o_i)
1.      curr_state = FALSE;
2.      while (true)
3.         command <-- receive(server);
4.         switch (command)
5.            case add_filter_constraint:
6.               Add new filter constraint to o_i;
7.               Stop;
8.            case delete_filter_constraint:
9.               Delete filter constraint from o_i;
10.              Stop;
11.        l_i <-- Sense_location();
12.        result <-- check_filter_constraints(l_i,curr_state);
13.        if (result == include_result)
14.           curr_state = TRUE;
15.           send_update(o_i,l_i,insert);
16.        else if (result == exclude_result)
17.           curr_state = FALSE;
18.           send_update(o_i,l_i,delete);
```

As we can see from Algorithm 1 (server side), after a query q is registered, the server will first collect the values. It will also calculate the filter constraints for q, and send them to the all objects (lines 1-4). During the lifetime of q (between time t_1 and t_2), the server will keep listening to status updates from the location objects, and update the uncertain database (lines 6-7). Depending on the update information, the received update will be used to remove(insert) the ID of o_i from(to) the query result (lines 8-11). After the query is finished, the filter constraints from all mobile objects involved are removed (Steps 12-13).

On the mobile object side, each object o_i first initializes its state of satisfying the query as "FALSE"(line 1). It then repeatedly listens to the commands from the server (lines 2-3). If the server command is to add or delete filter constraints (due to the arrival or completion of a CPRQ), it will do so accordingly (lines 4-10). Then, it will sense its location l_i (using GPS for example) (line 11), where l_i and its current state are checked against the filter constraint installed by the server (line 12). If the result of the checking indicates that o_i should be removed from the query result, o_i changes its current state and sends a message to the server about this request (lines 13-15). Similar treatments are done for including o_i in the query result (lines 16-18).

A number of interesting observations can be made for Algorithm 1:

- The amount of energy in transmitting the data depends on whether an update is required by the result of filter constraint checking (line 11). Hence, whether the algorithm is effective depends on the definition of the filter constraints.
- The server does not need to perform any numerical integration to recompute the probabilistic query result. It just needs to update the query result according to the command sent by the mobile object (lines 8-11).
- A filter constraint is easy for an object to be evaluated; an object does not need to perform numerical integration operations to compute its qualification probability at all. This is beneficial to mobile objects which may have weak computational power, and also saves the energy required for computation.
- The framework can be extended to support multiple queries that are running at the same time, by installing filter constraints derived for different queries, to the same mobile object.
- Algorithm 1 can be potentially extended to support other types of continuous probabilistic queries (e.g., nearest neighbor queries), by deriving different kinds of filters.

Let us now investigate the details of the filter constraints for CPRQ.

4.2   Deriving Filter Constraints

To illustrate how a filter constraint for a CPRQ should be generated, Figure 3(a) shows a range R and the uncertainty region of a mobile object A (the gray-colored circle). We assume that the probability threshold P of the query is equal to one. Observe that A is currently inside R – that is, A has a qualification probability of one. Thus, A is the included in the current query result. Now, if the next position of A is still inside R, then it is not necessary for the query result to be updated. Moreover, if A receives the information about R, then A can check by itself whether it needs to emit the update to the server. A "filter constraint" for A, when it is completely inside R, can then be defined as:

OUT $R \to$ Update

which reads: "When the uncertainty region of A touches R, report to the server." Thus, the server can first compute the above constraint and send it to A. As long as the current location of A does not satisfy the above condition, no update is reported.

We can apply the above observation to *any* value of P. Let us examine Figure 3 again, now assuming that $P = 0.7$. Suppose A continues to move towards the boundary of A, such that a fraction of more than 0.3 of its uncertainty region (shaded) lies outside R. At this time instant, A must report to the server, so that A can be removed from the query result. This is equivalent to setting the constraint "OUT $R' \to$ Update" for A, where R' is derived by using the maximum amount of A's uncertainty region allowed on the outside of R (equal to 0.3 of A's uncertainty region). In this figure, we represent R' as a rectangle. In general, however, R' may not be a rectangle. The derivation of the shape of R' is left for future study.

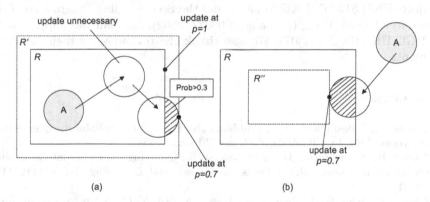

Fig. 3. Probabilistic filter constraints for an object (a) inside R, and (b) outside R

Figure 3(b) shows the case when A is currently outside R. For $p = 0.7$, the constraint "IN $R'' \to$ Update"' can be set for A. Notice that when the uncertainty region of A touches R'', it has a fraction of exactly 0.7 inside R. On receiving the update from A, the server should insert A to the query result. Observe that in this example, while R' is outside R, R'' is enclosed by R. Hence for for every CPRQ with a given p value, two constraints are required to handle both the two cases when the mobile object is inside or outside the query region.

From this example, we can see that once appropriate filter constraints have been set for an object, it can just decide whether reporting update is needed by checking against the filter constraints. This process can be done without computing its current qualification probability (which may require numerical integration). The scheme also alleviates the burden of the server, as once it receives an update from the object, it simply changees the query result, without recomputing the reporting object's probability.

5   Conclusions

In this paper, we study a protocol for processing continuous probabilistic queries over location data with uncertainty. Given a probability threshold P, we develop the probabilistic filter protocol, in order to reduce the transmission cost of a mobile object. Our rotocol also alleviates the computational effort of both the server and the mobile objects involved.

Our next step is to perform extensive simulation to validate our protocol under different settings. We also plan to extend the protocol to support the execution of multiple CPRQs that can be running concurrently. We will further consider new filter constraints to support other important location-based queries (e.g., nearest-neighbor queries).

Acknowledgments

Reynold Cheng was supported by the Research Grants Council of Hong Kong (Projects HKU 513307, HKU 513508), and the Seed Funding Programme of the University of Hong Kong (grant no.200808159002). Jin Jian was supported by the RGC HK (HKU 513507). We also thank the reviewers for their insightful comments.

References

1. Cheng, R., Kalashnikov, D.V., Prabhakar, S.: Evaluating probabilistic queries over imprecise data. In: SIGMOD 2003, pp. 551–562 (2003)
2. Cheng, R., Kalashnikov, D.V., Prabhakar, S.: Querying imprecise data in moving object environments. IEEE Trans. on Knowl. and Data Eng. 16(9), 1112–1127 (2004)
3. Cheng, R., Kao, B., Prabhakar, S., Kwan, A., Tu, Y.-C.: Adaptive stream filters for entity-based queries with non-value tolerance. In: VLDB 2005, pp. 37–48 (2005)
4. Cheng, R., Xia, Y., Prabhakar, S., Shah, R., Vitter, J.S.: Efficient indexing methods for probabilistic threshold queries over uncertain data. In: VLDB 2004, pp. 876–887 (2004)
5. Dalvi, N.N., Suciu, D.: Efficient query evaluation on probabilistic databases. In: VLDB 2004, pp. 864–875 (2004)
6. Gedik, B., Liu, L.: Mobieyes: Distributed processing of continuously moving queries on moving objects in a mobile system. In: Bertino, E., Christodoulakis, S., Plexousakis, D., Christophides, V., Koubarakis, M., Böhm, K., Ferrari, E. (eds.) EDBT 2004. LNCS, vol. 2992, pp. 67–87. Springer, Heidelberg (2004)
7. Gedik, B., Wu, K.-L., Yu, P.S.: Efficient construction of compact shedding filters for data stream processing. In: ICDE 2008, pp. 396–405 (2008)
8. Han, S., Chan, E., Cheng, R., Lam, K.-Y.: A statistics-based sensor selection scheme for continuous probabilistic queries in sensor networks. Real-Time Syst. 35(1), 33–58 (2007)
9. Hsueh, Y.-L., Zimmermann, R., Ku, W.-S.: Adaptive safe regions for continuous spatial queries over moving objects. In: Zhou, X., Yokota, H., Deng, K., Liu, Q. (eds.) DASFAA 2009. LNCS, vol. 5463, pp. 71–76. Springer, Heidelberg (2009)

10. Leonhardi, A., Rothermel, K.: Architecture of a large-scale location service. In: ICDCS 2002, p. 465. IEEE Computer Society, Washington (2002)
11. Ljosa, V., Singh, A.K.: Apla: Indexing arbitrary probability distributions. In: ICDE 2007, pp. 946–955 (2007)
12. Olston, C., Jiang, J., Widom, J.: Adaptive filters for continuous queries over distributed data streams. In: SIGMOD 2003, pp. 563–574. ACM, New York (2003)
13. Pfoser, D., Jensen, C.S.: Capturing the uncertainty of moving-object representations. In: Proceedings of the 6th International Symposium on Advances in Spatial Databases, pp. 111–132. Springer, London (1999)
14. Prabhakar, S., Xia, Y., Kalashnikov, D.V., Aref, W.G., Hambrusch, S.E.: Query indexing and velocity constrained indexing: Scalable techniques for continuous queries on moving objects. IEEE Trans. Comput. 51(10), 1124–1140 (2002)
15. Silberstein, A., Munagala, K., Yang, J.: Energy-efficient monitoring of extreme values in sensor networks. In: SIGMOD 2006, pp. 169–180. ACM, New York (2006)
16. Sistla, P.A., Wolfson, O., Chamberlain, S., Dao, S.: Querying the uncertain position of moving objects. In: Temporal Databases: Research and Practice, pp. 310–337. Springer, Heidelberg (1998)
17. Tao, Y., Cheng, R., Xiao, X., Ngai, W.K., Kao, B., Prabhakar, S.: Indexing multi-dimensional uncertain data with arbitrary probability density functions. In: VLDB 2005, pp. 922–933. VLDB Endowment (2005)
18. Wolfson, O., Sistla, A.P., Chamberlain, S., Yesha, Y.: Updating and querying databases that track mobile units. Distrib. Parallel Databases 7(3), 257–387 (1999)
19. Xiong, X., Mokbel, M.F., Aref, W.G.: Sea-cnn: Scalable processing of continuous k-nearest neighbor queries in spatio-temporal databases. In: ICDE 2005, pp. 643–654. IEEE Computer Society, Washington (2005)

Establishing Similarity across Multi-granular Topological–Relation Ontologies

Matthew P. Dube and Max J. Egenhofer

National Center for Geographic Information and Analysis
and
Department of Spatial Information Science and Engineering
University of Maine
Boardman Hall, Orono, ME 04469-5711, USA
matthew.dube@umit.maine.edu, max@spatial.maine.edu

Abstract. Within the Geospatial Semantic Web, selecting a different ontology for a spatial data set will enable that data's analysis in a different context. Analyses of multiple data sets, each based on a different ontology, require appropriate bridges across the ontologies. This paper focuses on establishing such a bridge across two topological-relation ontologies of different granularity—the standard *eight detailed* toplogical relations and *five coarse* topological relations. By mapping the conceptual neighborhood graphs onto a zonal representation, the different granularities are aligned spatially, yielding a reasoned approach to determining similarity values for the bridges across the two ontologies. A comparison with bridge lengths from an averaged model shows the better quality of zonal model.

1 Introduction

Geospatial ontologies that capture semantics of spatial information are paramount for the Geospatial Semantic Web [11, 18] in order to enable consistent spatial retrieval and Web-based geospatial services. The choice of an ontology provides an opportunity to analyze data in a different context or from a different perspective [16]. One of the major challenges that the Geospatial Semantic Web faces is when multiple spatial models need to be integrated, as in the case when data providers resort to using their own ontologies (i.e., data models and inference mechanisms) without providing explicit instruments to covert to other representations and reasoning methods. This paper deals with bridging across diverse spatial ontologies.

Some of the most advanced formalizations and agreements of spatial concepts are currently found in the domain of topological relations with sound formalisms [7, 13] . Since theses formalisms address abstract spatial models and inferences that permeate across all kinds of application domains they are typically considered essential ingredients for upper-level ontologies, such as OpenCyc [23] and the Standard Ontology for Ubiquitous and Pervasive Applications SOUPA [4].

While ontology alignment has been a stronghold in geospatial feature class consolidation [6], less effort has been put into the development of methods for bridging

K. Rothermel et al. (Eds.): QuaCon 2009, LNCS 5786, pp. 98–108, 2009.

between definitions of spatial relations, which differs from typical ontology alignments as the usual methods of lexical or structural comparisons [15] do not apply. Instead, a model-based integration is needed that exploits the inherent semantics of the underlying phenomena. This paper addresses the quality improvements of such alignments over a lexical or an *ad hoc* approach. Better knowledge about accommodating such spatial-relation ontologies across different granularities generalizes beyond the particular sets of relations studied and will lead to improved interoperability.

The purpose of this paper is to bridge spatial relation ontologies that address the same spatial concepts, but are defined at different granularities. The bridging should enable computations across the system divide. Such a system would allow for comparisons between coarse and detailed relations. Competing comparisons of coarse and detailed spatial relations may occur whenever two or more parties base their spatial analysis on different spatial relation ontologies. For example, from the perspective of a bank lending money to a potential new homeowner, the critical topological issue is whether the entire footprint of the building is contained on the land parcel, reflecting a coarse view of the topological relation *in* (disregarding whether parts of all of the building's boundary coincides with the land parcel's boundary). In a dispute with a neighbor about cleaning the house's gutters, however, a more detailed notion of inclusion applies as a boundary coincidence may require an easement, granting a right to step on the neighbor's property to maintain parts of the building. Comparisons across the two notions of inclusion require now conflict resolutions and mediations in the same way as the use of different spatial ontologies does for spatial entity classes [17].

While logical inferences over the detailed and coarse relations, implied by the relations' composition table [8], have been integrated into description logic languages [21] so that they are executable by automated reasoners [22], similarity reasoning over topological relations has been limited to relations of the same granularity [9]. Such similarity inferences need not only address the linkages among the most similar relations, but also require quantifications of such similarities in order to determine the cost for constraint relaxations. For purely topological reasoning this difference between each pair of most similar relations has been the unit within a complete set of the same granularity.

The remainder of the paper is structured as follows: Section 2 briefly reviews the sets of detailed and coarse topological relations and their conceptual neighborhood graphs. Section 3 analyzes how these graphs should be aligned. Section 4 introduces the zonal representation of the relations' neighborhoods and derives the lengths of the bridges across the relation ontologies. Section 5 analyzes the quality of the so established bridges and their lengths. The paper closes in Section 6 with conclusions and a discussion of future work.

2 Models for Topological Relations

The 9-intersection [13] and RCC, the Region-Connection Calculus [25], establish an ontology for topological spatial relations. The two methods yield equivalent results when the relations' range is restricted to 2-dimensional objects that are homeomorphic to regularized closed 2-disks (i.e., each region's closure is identical to the closure of the region's interior). For such simple regions, RCC and the 9-intersection identify a set of eight jointly exhaustive and pairwise disjoint relations, subsequently referred to by their 9-intersection terminology (Figure 1a).

Since the eight base relations form a relation algebra [28], disjunctions of these relations lead to *coarse* topological relations. A common generalization of the eight region relations is the RCC-5 abstraction [2], in which the relations *inside* and *coveredBy* are generalized to a single relation, here called IN, *contains* and *covers* generalize to IN^{-1}, *disjoint* and *meet* are combined into OUT, while the remaining two relations *overlap* and *equal* are mapped 1:1 onto OV and EQ, respectively (Figure 1b). These coarse relations have then the following interpretations: OUT means that both regions' interiors do not intersect; for OV regions share a portion of their interiors and exteriors with the opposing interior and exterior; for EQ the regions have coincident interiors; IN means that one region is a proper subset of the other; and IN^{-1} is converse to IN. Other mappings onto coarser topological relations—not considered here—consist of another domain over five relations with somewhat different mappings [20] or a domain with two relations (essentially distinguishing *equal* and its complement not *equal*).

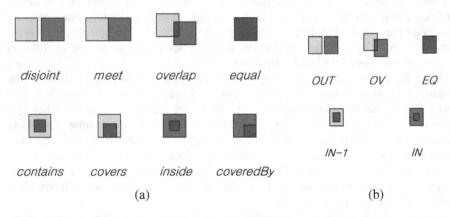

Fig. 1. Topological relations: (a) the eight detailed relations, (b) the five coarse relations

While the topological relations are on a nominal scale of measurements, the least amount of change needed to transform one configuration into another yields an order that is captured by the relations' conceptual neighborhood graph [7, 11]. Similar to Allen's [1] interval relations where different rationales for establishing neighborhoods lead to different neighborhood graphs [19], variations of the region relations' conceptual neighborhood graph can be found [5]. In the remainder of this paper only relation pairs captured by the A-neighborhood (based on translation, rotation and scaling free of any metric constraints on the involved regions, such as the regions' shapes) are considered (Figure 2a). In a similar way the eight coarse relations can be aligned forming their conceptual neighborhood graph (Figure 2b).

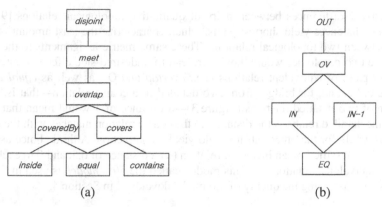

Fig. 2. The conceptual neighborhood graphs of (a) the eight detailed relations and (b) the five coarse topological relations

3 Aligning Detailed and Coarse Topological Relations

The conceptual neighborhoods form the foundation for qualitative similarity reasoning [3] as a relation's most similar relations correspond to that relation's neighbors in the graph, and increasingly less similar relations are neighbors of higher degrees. When reasoning about detailed *and* coarse topological relations, it is necessary to connect the two graphs. Such ontology alignment can be accomplished with a subset of the five binary semantic relations between two concepts [14]. It has two 1:1 mappings and six inclusions (Figure 3). These mappings build bridges in the graph between the detailed and the coarse relations (Figure 4). Computational similarity models require not only the association of most similar pairs, but also a quantification of the similarity [26] in order to support consistent reasoning.

$OV \equiv overlap \Rightarrow OV = overlap$ $EQ \equiv equal \Rightarrow EQ = equal$

$OUT \equiv disjoint \vee meet \Rightarrow OUT > disjoint$ $OUT \equiv disjoint \vee meet \Rightarrow OUT > meet$

$IN \equiv inside \vee coveredBy \Rightarrow IN > inside$ $IN \equiv inside \vee coveredBy \Rightarrow$
$$IN > coveredBy$$

$IN^{-1} \equiv contains \vee covers \Rightarrow IN^{-1} > contains$ $IN^{-1} \equiv contains \vee covers \Rightarrow$
$$IN^{-1} > covers$$

Fig. 3. Mapping detailed onto coarse topological relations using the semantic relations inclusion (<) and equivalence (=)

The mere alignment, however, offers no support for analytical operations such as similarity assessments involving detailed *and* coarse topological relations, which rely on the distances between the relations within the embedding of the neighborhood graph. For the isolated conceptual neighborhood graphs the distances along the graphs' edges are typically assumed to be of length 1 if only topological changes are considered. With this metric, path lengths can be established, which lay the foundation for determining quantitatively differences between pairs of qualitative topological relations [9]. Likewise these distances yield shortest paths, which characterize the least amount of differences between two topological relations. These same metric assignments to the bridges across the two ontologies would, however, lead to undesired side effects. First, the distance for pairs of equivalent relations (i.e., *overlap* and OV as well as *equal* and EQ) must be 0. Second, for bridges from two detailed to a coarse relation—that is, the disjunction of two detailed relations (Figure 3)—a distance of 1 would mean that for one half of the detailed relations the distance to the coarse relation matches with the detailed relations, while for the other half it would yield a contraction. A third choice assigns to each bridge classified by an inclusion relation (<) a distance of 0.5, and 0 to each bridge with an equivalence relation (=). This model, called the *50% model*, is used in Section 5 as a basis for assessing the quality of the model developed in Section 4.

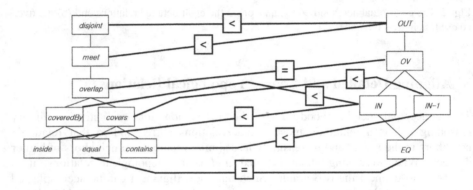

Fig. 4. The aligned conceptual neighborhood graph of the eight detailed relations and the five coarse topological relations with the semantic relations inclusion (<) and equivalence (=)

4   The Zonal Representation for the Neighborhood of Detailed and Coarse Topological Relations

We introduce for topological relations their *zonal representation*, which uses circles to capture a relation's range of influence and represents the similarities between region relations in the conceptual neighborhood graph (Figure 5a). Coarse relations that are not used as bridges need to fit somewhere into these circles. These circles are referred to by the name of the first relation from *overlap* contacted in the detailed topological relations. For example, any relation without an interior-interior intersection falls into the *meet* circle. Since *equal* is an intersection of *coveredBy* and *covers*, it falls on the intersecting point of these two circles. The next step connects *overlap* to the nodes for *meet*, *covers*, and *coveredBy*, and then to *disjoint*, *contains*, and *inside* (Figure 5b).

Each circle around *meet*, *coveredBy*, and *covers* has a radius of 1, therefore, all detailed relations maintain their initial similarity values from the conceptual neighborhood graph. This graph is actually just imposed over the circles. The purpose of the circles is to encapsulate the combinations among the coarse topological relations and to derive a distance measure for them. The zones that represent the coarse relations' combinations are shown in Figure 4c. Subsequently the *covers*-zone is used to exhibit the procedures for deriving the distances, since the remaining two zones follow suit under rotation.

Unlike the detailed topological relations *equal*, *overlap*, *contains*, and *covers*, the coarse relation IN^{-1} is represented as a zone rather than a node. It is desirable to come up with a node to represent the coarse relation so that a similarity measure can be assigned between it and any other relation in the mapping between coarse and detailed topological relations.

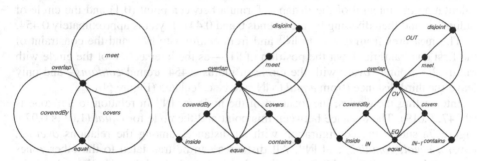

Fig. 5. (a) Zonal configuration for the detailed and coarse topological relations, (b) the conceptual neighborhood graph fused with the zones, and (c) the coarse relation constructs denoted on the zones.

The first distance to isolate is the distance between *covers* and IN^{-1}. This is rather straight forward as *covers* lies at the center of the unit circle. The double integral of the distance function over the unit circle from its center is $\pi/6$. Also the area of the first quadrant of the unit circle, representing the sector <*equal, covers, contains*> is $\pi/4$. The average distance from *covers* to any point in the zone denoted as IN^{-1} is then $(\pi/6)/(\pi/4)$, that is, 2/3 (Figure 6a).

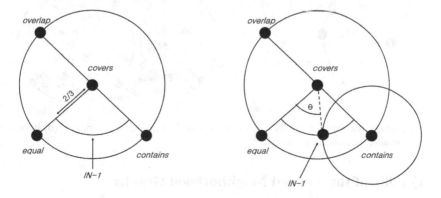

Fig. 6. (a) Arc representing IN^{-1} and (b) node for IN^{-1}

The next distance necessary is the distance from *contains* to IN^{-1}. The key to resolving this issue is how to place IN^{-1} with respect to its exemplars *contains*, *covers*, and *equal*. From a cognitive perspective, *contains* should be closer to IN^{-1} than *covers* or *equal*, because it is more of the prototype of the containment relation than *covers* or *equal*. Let *A* be the portion of the perimeter of the *covers* circle with radius 2/3 in the first quadrant, such that *x* (i.e., a point on the perimeter) is within 2/3 of the point (0,1), which represents *contains*. The desired value is the average distance from *contains* to any point in *A*. *A* is bounded by the x-coordinates 0 and $(2/3) \cdot \cos(\theta) = 0.441$, with $\theta \approx 0.848$ so that that the point on the perimeter is 2/3 away from (0,1). The distance between contains and IN^{-1} is computed by the mean value theorem, that is, calculating the integral of the distance formula between point (0,1) and the circle of radius 2/3 and then dividing by the bounds 0 and 0.441. It yields approximately 0.454.

The measures from *covers* to IN^{-1} and from *contains* to IN^{-1}, and the constraint of the first quadrant, tie down the position of IN^{-1}—as the intersection of the circle with radius 2/3 around *covers* with the circle of radius 0.454 around *contains*—and only leave the third distance (from *equal* to IN^{-1}) to be calculated (Figure 6b).

Intersecting the circles, the location of the node for IN^{-1} in relation to the zone is (0.247, 0.619). The distance between this point and the node for *equal* (1,0) is 0.975. Figure 7a shows the configuration with all distances among the relations *overlap*, *covers*, *equal*, *contains*, and IN^{-1}. All distances can be translated to the other zones (*meet* and *coveredBy*), yielding the aligned and quantified neighborhood graph for the combined detailed and coarse relations (Figure 7b). The three distances calculated under this method satisfy the triangle inequality; therefore, the aligned graph is a metric space.

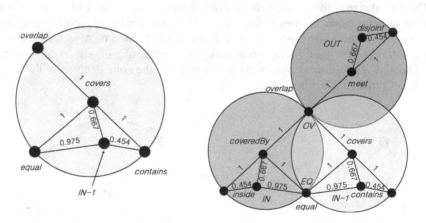

Fig. 7. Similarities between (a) IN^{-1} and *covers*, *equal*, *contains*, overlap, and (b) between the thirteen detailed and coarse relations

5   Quality of the Aligned Neighborhood Graphs

Several psychological findings guide the quality of such alignments. According to the *prototype theory*, people typically assign a dominant member of the superclass, becoming a salient image, and then compare all other objects against this prototype. Alternatively, the most central member of the superclass is selected as the prototype [27]. A competing theory is the *exemplar theory* of psychology, which starts with a basis of observations, called *exemplars*, that are known whether or not they belong to a set. Any subsequent objects then become compared to this group of objects to discern membership in a superclass [24]. Another related theory is Wittgenstein's [29] *graded sets*, which assign a hierarchy of dominance to members of a superclass, because not all members of the set are considered to give an adequate representation of the whole. In order to adequately perform such alignments, both the prototype theory and the exemplar theory must be taken into account by establishing a bounding set representing the coarse relations (exemplar theory), establishing a hierarchy of closest neighbors from the detailed up to the coarse relations (graded set), and from this graded set, discerning the prototypical member (prototype theory). Bounding relations from the detailed topological relations already have been established, providing a methodical basis by which to formulate scene similarities between other systems with inherent bridges or other systems, which are essentially disjunctions of members of the detailed relations.

The complete set of distances for the Cartesian product of detailed and coarse relations (Figure 8) reveals a number of properties of the aligned similarity graphs:

- The shortest distance between any two different relations is 0.454, which has three occurrences (always between a detailed and a coarse relation and, therefore, part of the alignment). This value is less than the standard distance between two coarse or two detailed relations in the isolated neighborhood graphs.
- The longest distance between two relations is 4, the same as the longest distance in the isolated neighborhood graph of detailed relations; therefore, the alignment has no effect on the longest shortest path among the detailed relations.
- The distances of the bridges from detailed to coarse relations range from 0 (since *equal* = EQ and *overlap* = OV) to 0.667. The upper bound is considerably less than the standard distances between neighboring pairs in the isolated graph of detailed relations and the isolated graph of coarse relations. Apparently a homogeneous choice of 1 as the length of all edges of the aligned graphs would be counter to the geometric arguments that emerged out of the zonal model.
- The post-alignment distances among the eight detailed relations coincide with their corresponding pre-alignment distances in 54 out of 64 cases (which corresponds to an 82% agreement when accounting for implied values of converse cases and the distance between each relation and itself). For the five detailed relation pairs that are impacted by the alignment (*equal–inside, contains–covers, contains–inside, equal–contains, inside–coveredBy*) the post-alignment distance is shorter than the pre-alignment distance.
- The post-alignment distances among the five coarse relations coincide with their corresponding pre-alignment distances in 7 out of 25 cases, that is, only one distance—between OV and EQ—remains unchanged when accounting for implied values of converse cases and the distance between each relation and itself. Three post-alignment distances (and their reverse distances) are shorter than their corresponding pre-alignment distances—IN–IN^{-1}, IN–EQ, and IN^{-1}–EQ—whereas the remaining six relation pairs—OUT-OV, OUT-IN, OUT-IN^{-1}, OUT-EQ, OV-IN, and OV-IN^{-1}—show increases in their post-alignment distances.

	disjoint	meet	overlap	coveredBy	covers	inside	contains	equal	OUT	OV	IN	IN⁻¹	EQ
disjoint	0	1	2	3	3	4	4	4	A	2	3+B	3+B	4
meet	1	0	1	2	2	3	3	3	B	1	2+B	2+B	3
overlap	2	1	0	1	1	2	2	2	1+B	0	1+B	1+B	2
coveredBy	3	2	1	0	2	1	1+A+C	1	2+B	1	B	1+C	1
covers	3	2	1	2	0	1+A+C	1	1	2+B	1	1+C	B	1
inside	4	3	2	1	1+A+C	0	2A+2C	A+C	3+B	2	A	+2C	+C
contains	4	3	2	1+A+C	1	2A+2C	0	A+C	3+B	2	+2C	A	+C
equal	4	3	2	1	1	A+C	A+C	0	3+B	2	C	C	0
OUT	A	B	1+B	2+B	2+B	3+B	3+B	3+B	0	1+B	2+B	+2B	+B
OV	2	1	0	1	1	2	2	2	1+B	0	1+B	1+B	2
IN	3+B	2+B	1+B	B	1+C	A	A+2C	C	2+2B	1+B	0	2C	C
IN⁻¹	3+B	2+B	1+B	1+C	B	A+2C	A	C	2+2B	1+B	2C	0	C
EQ	4	3	2	1	1	A+C	A+C	0	3+B	2	C	C	0

Fig. 8. Dissimilarities between all pairs of coarse and detailed topological relations (A=0.454; B=0.667; C= 0.975)

As a quantitative base for the assessment of the aligned neighborhood graph we investigate the relations along the neat line of the two isolated graphs and their bridges. Two bridges create two paths from the detailed to the coarse relation—one by first crossing the bridge, followed by a transition to the coarse target relations and the other by first making the transition to the linked detailed relation, followed by crossing the bridge from there to the coarse relation. More similar path lengths are considered as an indicator of better alignments. We selected three such pairs of bridges and compared the path lengths for the zonal model (Section 4) and the 50%-model

Path	Zonal Model		50%-Model	
	path length	path length diff.	path length	path length diff.
disjoint-OUT-OV	2.121	0.121	1.5	0.5
disjoint-overlap-OV	2		2	
overlap-OV-IN	1.667	0.787	1	1.5
overlap-inside-IN	2.454		2.5	
overlap-OV-EQ	2	0	2	0
overlap-equal-EQ	2		2	

Fig. 9. Path length differences in the zonal and 50%-model

(Section 3). In two cases the zonal model yielded smaller differences than the 50%-model, and in one case both featured the same path length differences, which supports the zonal model's quality alignment (Figure 9).

6 Conclusions

We developed a method to attach two topological-relation ontologies of different granularities so that the bridges across the ontologies carry lengths that are compatible with the relations' conceptual neighborhood graphs. The comparison with other distance choices showed the pointed advantage of the zonal model. A number of open questions remain for future work:

- How does the method perform if each neighborhood graph is first normalized by longest shortest path prior to bridging?
- How does the coarse relation alignment differ from attaching a somewhat differently defined set of five coarse topological relations [20]?
- How would an even coarser set of only two relations (*same* and *different*) behave in this process, particularly with respect to transitivity?

Acknowledgments

This work was partially supported by the National Geospatial-Intelligence Agency under grant number NMA201-01-1-2003 and NMA401-02-1-2009.

References

[1] Allen, J.: Spatial Reasoning with Propositional Logics. In: Doyle, J., Sandewall, E., Torasso, P. (eds.) 4th International Conference on Principles of Knowledge Representation and Reasoning (KR 1994), pp. 51–62. Morgan Kaufmann, San Francisco (1983/1994)

[2] Bennett, B.: Spatial Reasoning with Propositional Logics. In: Doyle, J., Sandewall, E., Torasso, P. (eds.) 4th International Conference on Principles of Knowledge Representation and Reasoning (KR 1994), pp. 51–62. Morgan Kaufmann, San Francisco (1994)

[3] Bruns, H.T., Egenhofer, M.: Similarity of Spatial Scenes. In: Kraak, J.-M., Molenaar, M. (eds.) Seventh International Symposium on Spatial Data Handling, pp. 173–184. Delft, Taylor & Francis, The Netherlands, London (1996)

[4] Chen, H., Finin, T., Joshi, A.: The SOUPA Ontology for Pervasive Computing. In: Tamma, V., Cranefield, S., Finin, T. (eds.) Ontologies for Agents: Theory and Experiences. Springer, Heidelberg (2005)

[5] Cohn, A., Gooday, J., Bennett, B.: A Comparison of Structures in Spatial and Temporal Logics. In: Casati, R., Smith, B., White, G. (eds.) Philosophy and the Cognitive Sciences, Hödler-Pichler-Tempsky, Vienna, Austria, pp. 409–422 (1994)

[6] Cruz, I., Sunna, W.: Structural Alignment Methods with Applications to Geospatial Ontologies. Transactions in GIS 12(6), 683–711 (2008)

[7] Cui, Z., Cohn, A., Randell, D.: Qualitative Simulation Based on a Logical Formalism of Space and Time. In: Swartout, W. (ed.) AAAI-92—10th National Conference on Artificial Intelligence, pp. 679–684 (1992)

[8] Egenhofer, M.: Reasoning about Binary Topological Relations. In: Günther, O., Schek, H.-J. (eds.) SSD 1991. LNCS, vol. 525, pp. 143–160. Springer, Heidelberg (1991)

[9] Egenhofer, M.: Query Processing in Spatial-Query-by-Sketch. Journal of Visual Languages and Computing 8(4), 403–424 (1997)
[10] Egenhofer, M.: Toward the Semantic Geospatial Web. In: Voisard, A., Chen, S.-C. (eds.) ACM-GIS 2002, McLean, VA, November 2002, pp. 1–4 (2002)
[11] Egenhofer, M., Al-Taha, K.: Reasoning about Gradual Changes of Topological Relationships. In: Frank, A.U., Formentini, U., Campari, I. (eds.) GIS 1992. LNCS, vol. 639, pp. 196–219. Springer, Heidelberg (1992)
[12] Egenhofer, M.J., Franzosa, R.D.: Point-Set Topological Relations. International Journal for Geographical Information Systems 5(2), 161–174 (1991)
[13] Egenhofer, M., Herring, J.: Categorizing Binary Topological Relations Between Regions, Lines, and Points in Geographic Databases. Technical Report, Department of Surveying Engineering. University of Maine (1990)
[14] Euzenat, J.: Algebras of Ontology Alignment Relations. In: Sheth, A.P., Staab, S., Dean, M., Paolucci, M., Maynard, D., Finin, T., Thirunarayan, K. (eds.) ISWC 2008. LNCS, vol. 5318, pp. 387–402. Springer, Heidelberg (2008)
[15] Euzenat, J., Shvaiko, P.: Ontology Matching. Springer, Heidelberg (2007)
[16] Fonseca, F., Egenhofer, M.: Ontology-Driven Geographic Information Systems. In: Bauzer Medeiros, C. (ed.) ACM-GIS 1999—Seventh Symposium on Advances in Geographic Information Systems, pp. 14–19. ACM Press, New York (1999)
[17] Fonseca, F., Egenhofer, M., Agouris, P., Câmara, G.: Using Ontologies for Integrated Geographic Information Systems. Transactions in GIS 6(3), 231–257 (2002)
[18] Fonseca, F., Sheth, A.: The Geospatial Semantic Web, UCGIS Research Priority (2002), http://www.ucgis.org/priorities/research/2002researchPDF/shortterm/e_geosemantic_web.pdf
[19] Freksa, C.: Temporal Reasoning based on Semi-Intervals. Artificial Intelligence 54(1-2), 199–227 (1992)
[20] Grigni, M., Papadias, D., Papadimitriou, C.: Topological Inference. In: Fourteenth International Joint Conference on Artificial Intelligence, pp. 901–907 (1995)
[21] Haarslev, V., Lutz, C., Möller, R.: A Description Logic with Concrete Domains and a Role-Forming Predicate Operator. Journal of Logic and Computation 9(3), 351–384 (1999)
[22] Haarslev, V., Möller, R.: RACER System Description. In: Goré, R.P., Leitsch, A., Nipkow, T. (eds.) IJCAR 2001. LNCS (LNAI), vol. 2083, pp. 701–706. Springer, Heidelberg (2001)
[23] Lenat, D., Guha, R.V.: Building Large Knowledge-Based Systems: Representation and Inference in the Cyc Project, February 1990. Addison-Wesley, Reading (1990)
[24] Medin, D., Schaffer, M.: Context Theory of Classification Learning. Psychological Review 85(3), 207–238 (1978)
[25] Randell, D., Cui, Z., Cohn, A.: A Spatial Logic based on Regions and Connection. In: KR92: Principles of Knowledge Representation and Reasoning: Proceedings of the Third International Conference, San Mateo, CA, pp. 165–176 (1992)
[26] Rodríguez, A., Egenhofer, M.: Determining Semantic Similarity Among Entity Classes from Different Ontologies. IEEE Transactions on Knowledge and Data Engineering 15(2), 442–456 (2003)
[27] Rosch, E.: Natural Categories. Cognitive Psychology 4(3), 328–350 (1973)
[28] Smith, T., Park, K.: Algebraic Approach to Spatial Reasoning. International Journal of Geographical Information Systems 6(3), 177–192 (1992)
[29] Wittgenstein, L.: Philosophical Investigations. Blackwell, Hoboken (1953)

UDS: Sustaining Quality of Context Using Uninterruptible Data Supply System

Naoya Namatame[1], Jin Nakazawa[2], Kazunori Takashio[1,2],
and Hideyuki Tokuda[1,2]

[1] Graduate School of Media and Governance, Keio University, Kanagawa, Japan
namachan@ht.sfc.keio.ac.jp
[2] Faculty of Environmental Information, Keio University, Kanagawa, Japan
{hxt,kaz,jin}@ht.sfc.keio.ac.jp

Abstract. Context mining algorithms from sensor data have been researched and successful results have been shown. However, since these existing works are focused on improving the accuracy of context mining, they are established on the assumption that they can acquire a complete set of necessary data. Therefore, the context mining algorithms do not work sufficiently since the data drops easily in the reality. In this paper, to cope with this problem, we propose a middleware named UDS (Uninterruptible Data Supply System). The system compensates the missing data, creates virtually complete dataset and provides upper layer applications. Applications operating over UDS can work sufficiently with some data actually missing. We have defined two types of characteristic data deficit patterns and created a robust model for both patterns utilizing Bayesian Network. In the evaluation, we show UDS can sustain the quality of context over 80% with 40% data missing.

Keywords: Data Compensation, Context Inference, Reliable System.

1 Introduction

"Context", generally in ubiquitous computing, can be explained as real space information that can be a useful criterion for an application to decide a service to provide. Applications that utilize them are called "Context-Aware Application" and many of them are proposed [2,9,10,11]. Context is usually described in abstract ways, for example, the room is "used for meeting" or "used for lunch". Applications have to infer the context from sensor data since it is impossible to acquire this type of abstract contextual information directly from raw data of sensors. Therefore, accuracy of context inference result, which can also be defined as "Quality of Context", is significantly important in building a practical context aware application. Many of research on data collection is also proposed [8,16,17].

Many researchers have been working on improving accuracy of context mining and showing successful results [4,5,6]. However, context aware systems have not yet been appeared in real life. One of the reasons for this is that many of

K. Rothermel et al. (Eds.): QuaCon 2009, LNCS 5786, pp. 109–119, 2009.
© Springer-Verlag Berlin Heidelberg 2009

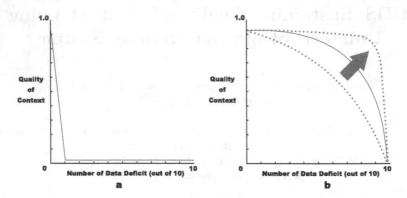

Fig. 1. Relationship between the quality of context and the number of data deficit

researches focus on the method to improve the quality of context and do not care about **sustaining** the quality of context in a real use. Sensors, which are the source of the context, may exhaust battery or failed to receive packets by radio wave interference. There are many factors that an application can not receive a complete set of sensor data.

These proposed algorithms are not guaranteed to work sufficiently with lack of sensor data. Some algorithms such as Neural Network expect all the necessary data to be set as input. They suddenly become useless as the graph indicates in Fig.1 with a lack of data. The graph represents that the quality of context decline straight to 0 with a small part of data missing even if an algorithm normally infers context with extremely high quality.

On the other hand, Fig.1b shows the ideal transition of the quality of context. In the graph, the quality of context stays high until the number of data deficit gets a few and then starts declining gradually. The context aware system can work sufficiently until the number of data deficit becomes high in this way.

This fragility of quality of context shown in Fig.1a can be one of major problems that not many context-aware applications appear on our life although the notion of context-aware application is proposed 15 years ago in 1994[1].

To cope with this problem, we propose "Uninterruptible Data Supplying System (UDS)", a sensor data compensation middle ware utilizing Bayesian Network to sustain the quality of context with fragile data sources. Similar to Uninterruptible Power Supplying System (UPS), UDS infers and compensates missing data and supplies virtual complete data set. It can be adopted to existing works without changing them since the approach working different layer from existing context inference algorithms.

The rest of the paper is organized as follows. In the second section, we explain the related works of UDS and describe the difference between UDS and them. Then in the third section, we explain UDS system and enabling technologies. In the fourth section, the evaluation result of UDS is described. After that, discussion section follows and at last section, future work and conclusion has been stated.

2   Related Works

Having the same motivation, Murao et al[3] proposed a data complementing method considering breakdowns of wearable sensors. They have also developed a device named CLAD that controls and observes power supply and state of the sensors. By utilizing both technologies, they proposed a way to create a dependable wearable sensor network environment. As for the data complementing method, they process received incomplete data pattern with k-Nearest Neighbor Algorithm utilizing stored complete data pattern histories.

As they target on wired wearable sensor environment, our research has a different target environment. Our target is a sensor embedded space, which is an environment that many objects, for example desks, chairs, and lights, have small sensors on them. Packet losses caused by radio interferences will appear more often than sensor breakdown in wireless sensor networks. In this paper, we also recognize radio interferences as another threat, and propose a method to improve the precision for complementing those temporal packet losses. Also, by utilizing probabilistic model, our system do not need to store many raw patterns of peripheral sensor data.

3   UDS: Uninterruptible Data Supply System

To sustain the quality of context in a fragile data source environment, we proposed UDS, which stands for Uninterruptible Data Supply System. UDS complements missing data that the system is supposed to receive. Then, UDS create a virtual complete dataset and supply it to upper layer context aware applications.

System structure image is shown in Fig.2. Normally the application acquires the sensor data directly from UART or some type of bass connected to a sink node. With UDS, first UDS, which works as a middleware, collects data from the sink node and then provides data to the application with complementing lack of necessary sensor data. In this way, they can use UDS without changing them regardless of the algorithm that the existing applications have.

3.1   Bayesian Network Data Compensation

We have adopted Bayesian Network inference algorithm to infer the sensor data that the system could not observe. Bayesian Network algorithm is well

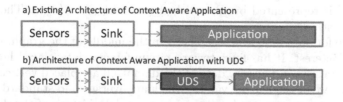

Fig. 2. Architecture of context aware system with and without UDS

known as the algorithm that can handle three kinds of uncertainties, that are from "IGNORANCE", "PHYSICAL RANDOMNESS or INDETERMININS" and "VAGUENESS"[7]. Since we sometimes DO NOT KNOW what the sensor data is, the lack of data can be regarded as the uncertainty from ignorance. Using this aspect of view, we assume that Bayesian Network algorithm can be applied to data compensation in sensor network and contribute to sustain the quality of context.

3.2   Robust Inference Model

To create Bayesian Network inference model suitable for sensor data compensation, we focus on two types of data deficit patterns. One is "Temporal Data Deficit" and the other is "Constant Data Deficit".

- Temporal Data Deficit
  Temporal Data Deficit can be seen when radio wave collisions or interferences are occurred. The system fails to receive just for a packet or a few. But after that, packets are normally received.

- Constant Data Deficit
  Constant data Deficit can be seen when a node breaks down or exhausts its battery. The node stops sending packet after a node breaks down or exhausts its battery,. Therefore the system can not receive the packet from the node constantly after that.

In order to be tolerant for both types of data deficit, we have utilized data from circumference nodes and history data of the node itself. First, the reason for utilizing data from circumference nodes is that a series of data from sensors placed in the same environment creates patterns and they can be utilized to back calculate a part of the dataset. The data patterns of circumference nodes contribute to infer temporal data deficit. However, data compensation precision is expected to decline dramatically when the number of constant data deficit increases with only data of circumference nodes. To cover this situation, history data of the node itself is needed. History data of the node itself is useful to infer the state of itself since some of sensor data values transit based on their previous state. The inference model becomes more tolerant to the number of constantly unavailable circumference data by adding this information to the model. The model created with these data source is shown in Fig.3. Sensor data at a certain time period is represented by $D_t = (d_{1,t}), (d_{2,t}), (d_{3,t}), ..., (d_{n,t})$. The model is created for each $d_{i,t}$.

The node for data $d_{i,t}$, which is placed in the middle of the graph, represents the data of interest. It has links toward circumference data placed horizontally on the bottom area and history data of the data itself are placed vertically on the right hand. Two of the feature values, which are average and standard deviation of the $d_{i,t}$ shown as $mu_{i,t}$ and $delta_{i,t}$, represent data history of the deficit data itself. Since there is no need for hidden layer for this inference, we adopted Naive

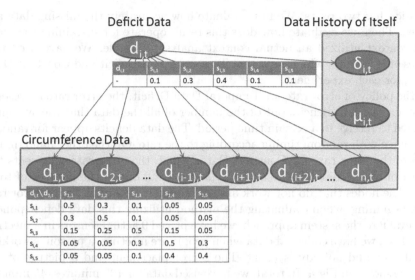

Fig. 3. Bayesian network model for data compensation

Bayes model. Users can create model without any specific knowledge and the model can be created semi-automatically in this way.

The node for data $d_{i,t}$ has a probability table of the occurrence ratio for each its state $S_i = s_{i,1}, s_{i,2}...s_{i,m}$. The child nodes, which have a link from the node for data $d_{i,t}$ in the graph, have a conditional probability table of the occurrence ratio of their states which are conditioned by the each state of S_i. These tables are set by probabilities calculated from the complete set of data collected beforehand.

3.3 Quick Responsive Algorithms

In order to shorten the calculation time for data complement, we have adopted a type of Approximate Inference of Bayesian Network named Likelihood Weighting. When utilizing exact inference in our case, calculating 3 unobserved data from a model which has 10 data types with 5 states for each, the system needs to calculate $5^{10} times 3 = 29296875$ patterns of probability. This number may be reasonable but the number increases exponentially to the number of total data. Exact inference is not appropriate when utilizing this method for environment with hundreds of sensors installed. Approximate inference calculates the probability with stochastic simulation, so that they can reach to the approximate value for the exact logical probability. For the method to be scalable, we have adopted Approximate Inference.

4   Evaluation

In this section, we will verify that Bayesian Network algorithm and the proposed inference model is effective to data compensation in sensor networks. The evaluation has two parts. We first evaluate how accurately the missing data are inferred. Then, we evaluate how does this result operate for sustaining the quality of context utilized an actual context analysis example. We have used two policies of error rate, which represent Temporal Data Deficit and Constant Data Deficit, for each experiment.

In the policy of error rate for Temporal Data Deficit, the error rate represents the number of data deficits out of the number of all the data that the system is supposed to receive in a certain time period. The data deficits appear for random data source with random timing according to the rate. On the other hand, in the policy of the error rate for Constant Data Deficit, the error rate represents the number of nodes that do not send any packet at all out of the number of total nodes. The nodes that do not work are set randomly according to the error rate at the beginning, when evaluating the method utilizing this data deficit policy.

To evaluate the system approach, we have placed 9 actual sensors in a meeting room. Then we have collected data when people are meeting, a person is working individually, and nobody is using. The sensor placement and a picture of the room are shown in Fig.4. In total, we have used data for 135 minutes, 45 minutes for each context, and all sensors are set to send data for every 10 seconds.

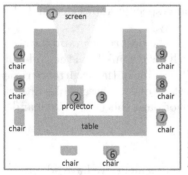

Fig. 4. Sensor placement

The data types we used are 1) luminance, 2) temperature and 3) acceleration in x-axis. As for data for acceleration in x-axis, they represent the absolute value for maximum acceleration detected in one sensing cycle. Each data is quantized equally into 10 levels. We have set granularity of quantization 10lv for luminance, $5Ccirc$ for temperature, and 0.01G for acceleration. The target dataset has data for 30 minutes, which are chosen from the three situations, 10 minutes for each. In the evaluation process, we evaluate how accurate the compensation results are matched to the actual ones.

4.1   Performance of Data Compensation

First, we focus only on the performance of the data compensation method that we proposed. We have executed simulations for 10 times and evaluated the

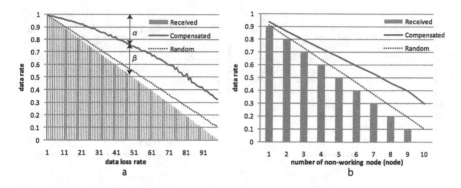

Fig. 5. Precision of data compensation with a)temporal and b)constant data deficit

accuracy of data compensation. In Fig.5, a bar chart represents the number of data that are actually received. A line chart with solid line represents the number of data that are corresponding to the reality, which is consist of 1) the data that are actually received and 2) the data that are missing but correctly inferred with our algorithm. And a line chart with dotted line represents the same as solid line but the data are randomly inferred. Fig.5a shows the result of compensation with temporal data deficit occurance pattern starting from 1% of data loss up to 100%. Fig.5b shows the result of the evaluation with Constant Data Deficit policy starting from 1 node down to all nodes down utilizing the same dataset from the last evaluation.

For both results, the data compensation algorithm operates properly. Compared to the performance of random compensation, the performance of our compensation method operates better. Using α and β shown in Fig.5a, data compensation precision can be defined as $accuracy = \beta/(\alpha + \beta)$. $\alpha + \beta$ represents total missing data and α represents the part which UDS failed to infer, and β represents the part which UDS inferred correctly. Each accuracy of the data compensation is shown in Table.1. Considering random compensation precision is around 10% at most of the times, both precisions are at least approximately 3 times, at most 6 times better than random inference.

4.2 Sustainability of Quality of Context

Since it is not sufficient to say the result is appropriate enough to sustain the quality of context, we have tested how this system operates to sustain the quality

Table 1. Precision of data compensation for both temporal and constant data deficit

data loss rate (%)	0.1	0.2	0.3	0.4	0.5	0.6	0.7	0.8	0.9
temporal	0.674	0.624	0.580	0.561	0.492	0.484	0.429	0.387	0.360
constant	0.382	0.321	0.319	0.326	0.325	0.321	0.334	0.320	0.329

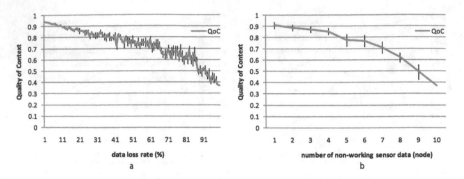

Fig. 6. Sustainability for quality of context with a)temporal and b)constant data deficit

of context in an actual context aware system. To evaluate this, we have developed a context extraction system independently with neural network and tried to extract the three contexts of the room. For the input value of the neural network, we have utilized the number of chairs that have been detected movement within last 3 minutes, temperature of projector, luminance of the room and screen. We have set this experimental settings since preciseness of the characteristic values which are utilized in the inference will declines as the number of available data decreases. In addition to that Neural Network needs complete dataset to execute its inference.

As a result of context extraction with complete dataset, the system could recognize the context that the room is used for meeting with 100.0% of precision, used by an individual with 91.11% and used by no one for 91.11%. In total, the quality of context was 94.4%. We have used this context extraction setting for the evaluation and executed simulations for 10 times and calculate their average and standard deviation same as the last time.

Sustainbility for the quality of context have turned out to be sufficient for both temporal and constant data deficit. The graph is shown in Fig.6. The both lines draw a moderate curve and decline to 37.3% starting from maximum value 94.4%. In both temporal and constant data deficit patterns, UDS could hold the quality of context above 80% until the amount of missing data gets over 40%. This indicates that the system can constantly keep operation without any problem in the environment with some packet losses observed. Also UDS gives system administrators enough time to replace broken sensors or change batteries of them. Detailed result is shown in Table2.

Table 2. Precision of context extraction with data UDS for both temporal and constant data deficit

data loss rate (%)	0.1	0.2	0.3	0.4	0.5	0.6	0.7	0.8	0.9	1.0
temporal	0.916	0.881	0.808	0.807	0.746	0.739	0.646	0.592	0.472	0.372
constant	0.911	0.889	0.872	0.853	0.776	0.768	0.710	0.621	0.492	0.371

5   Discussion

We have come up with three important discussions through developing and examining UDS even though we had a successful result from the system evaluation. These observations are directly linked to our future work.

5.1   Inference of Reality or Imitation of Reality

We have conducted this research with a hypothesis that the sustainability of the quality of context increases in promotion to the precision of data compensation. Therefore, we implemented the system to choose the state whose occurrence probability is calculated at most by Bayesian Network in a certain condition as an answer of inference result. In this way, the precision of data compensation should become the highest. However, suppose a person is sitting on a chair and the chair moves once in a while in a certain condition as shown in Fig.7a. In our method, the system infers the chair never moves as shown in Fig.7b with this condition. This data compensation result indicates that noone is sitting on the chair. The data compensation precision is higher in this way for sure, but the result is totally different from the reality. There is a paradox that the intension to calculate result as precise as possible to the reality leads to the result which is totally different from the reality. To sustain the quality of context, especially abstract ones, it might be a better way that we do not become obsessive about preciseness and try to just imitate reality as Fig.7c. For example, instead of choosing most logical answer, choosing the result randomly based on the probability which Bayesian Network calculates.

5.2   Dependable or Undependable

Now, the UDS that we have implemented supplies the complete dataset as if all the data are available regardless of the amount of missing data. However, there is a question that the system should supply data in any condition. To stop supplying data if necessary may be another important function. If compensated complete dataset is supplied constantly in any condition, upper layer applications can work constantly even if no data is available. This can cause problems in some applications although this fact is a merit of this system. For example, a nursing assisting application running on the UDS should not supply dataset with low dependability since the application do not allow many mistakes, In such case, administrator of the application may want to notify when the data dependency level becomes lower, and stop supplying all data if the dependency level gets lower than a threshold. We should discuss and define UDS's data supplement policy such as "Dependable" or "Best Effort". At the same time, the calculation algorithm for dependableness needs to be considered.

Fig. 7. Logical answer and imitation of reality

5.3 Level of Abstraction for Target Context

We have proposed a new category of middlewares for Context-Aware applications. This time, we have implemented UDS specifically for Context-Aware applications which manage highly abstract context such as "people are having meeting". These contexts tend to have time span or multiple definitions. For example, meetings usually last more than 10 minutes, and the number of the participants differ every time. It is not always necessary to compensate missing data exactly since upper ayer context inference algorithm alleviates the small errors in infering this kind of abstract context. This allows the implementation of UDS operates properly, although the data compensation accuracy is low when looking at the data compensation result of the method which we have proposed. This means the implementation of UDS can not be adopted to highly concrete context such as human movement tracking [12,13] since human location do not have time span, and the location of $time_t$. Image information acquired by camera is also out of support [14,15]. It is necessary to decide a level of abstraction for target context which this implementation of UDS can support.

6 Conclusion and Future Work

In this paper, we regarded the term quality of context as the context inference accuracy from sensor data and brought out a strong necessity of sensor data compensation to sustain the quality of context. Then, we have proposed and implemented a new concept of sensor network middleware named UDS, which stands for Uninterruptible Data Supply System. The system complements missing data from wireless sensors and keep supplying virtual complete dataset to upper layer applications. We have adopted Bayesian Network inference algorithm and showed its performances which are precision of data compensation and sustainability of the quality of context. For the future work, we are considering two approaches to improve sustainability for the quality of context. One is to keep working on improving data compensation precision by reconsideration of the inference model or the algorithm.

The other approach is, as we have mentioned in the discussion section, disregard data compensation precision and try to imitate real situation. Also, we have to evaluate the UDS under variety of applications and make sure the algorithm and the model are versatile enough to adopt a wide range of applications.

Acknowledgment

This work was supported by National Institute of Communication Technoligy (NICT) as a part of "Dynamic Network Project".

References

1. Bill, S., Norman, A., Roy, W.: Context-aware computing applications. In: Proceedings of the Workshop on Mobile Computing Systems and Applications, pp. 85–90. IEEE Computer Society, Los Alamitos

2. Hans, W.G., Michael, B., Holger, K.: The Media Cup: Awareness Technology Embedded in an Everyday Object. In: Gellersen, H.-W. (ed.) HUC 1999. LNCS, vol. 1707, pp. 308–310. Springer, Heidelberg (1999)
3. Murao, K., Takegawa, Y., Terada, T., Nishio, S.: A Sensed Data Complementing Method Considering Breakdowns of Wearable Sensors for Wearable Computing Systems. In: DEWS 2007 (2007)
4. Aipperspach, R., Cohen, E., Canny, J.: Modeling human behavior from simple sensors in the home. In: Fishkin, K.P., Schiele, B., Nixon, P., Quigley, A. (eds.) PERVASIVE 2006. LNCS, vol. 3968, pp. 337–348. Springer, Heidelberg (2006)
5. Intille, S.S., Larson, K., Tapia, E.M., Beaudin, J.S., Kaushik, P., Nawyn, J., Rockinson, R.: Using a live-in laboratory for ubiquitous computing research. In: Fishkin, K.P., Schiele, B., Nixon, P., Quigley, A. (eds.) PERVASIVE 2006. LNCS, vol. 3968, pp. 349–365. Springer, Heidelberg (2006)
6. Lester, J., Choudhury, T., Borriello, G.: A practical approach to recognizing physical activities. In: Fishkin, K.P., Schiele, B., Nixon, P., Quigley, A. (eds.) PERVASIVE 2006. LNCS, vol. 3968, pp. 1–16. Springer, Heidelberg (2006)
7. Korb, K., Nicholson, A.: Bayesian Artificial Intelligence. In: CRC Pr I Llc (2003)
8. Tapia, E.M., Intille, S.S., Lopez, L., Larson, K.: The Design of a Portable Kit of Wireless Sensors for Naturalistic Data Collection. In: Fishkin, K.P., Schiele, B., Nixon, P., Quigley, A. (eds.) PERVASIVE 2006. LNCS, vol. 3968, pp. 117–134. Springer, Heidelberg (2006)
9. Barkhuus, L., Dourish, P.: Everyday Encounters with Context-Aware Computing in a Campus Environment. In: Davies, N., Mynatt, E.D., Siio, I. (eds.) UbiComp 2004. LNCS, vol. 3205, pp. 232–249. Springer, Heidelberg (2004)
10. Cheverst, K., Davies, N., Mitchell, K., Friday, A., Efstratiou, C.: Developing a Context-Aware Electronic Tourist Guide: Some Issues and Experiences. In: Proceedings of MOBICOM 2000. ACM Press, New York (2000)
11. Burrell, J., Gay, G.K., Kubo, K., Farina, N.: Context-Aware Computing: A Test Case. In: Borriello, G., Holmquist, L.E. (eds.) UbiComp 2002. LNCS, vol. 2498, p. 1. Springer, Heidelberg (2002)
12. Hori, T., Nishida, Y., Aizawa, H., Murakami, S., Mizoguchi, H.: Distributed Sensor Network for a Home for the Aged. In: 2004 IEEE International Conference on Systems, Man and Cybernetics, pp. 1577–1582 (2004)
13. Standord, V.: Using Pervasive Computing to Deliver Elder Care. IEEE Pervasive Computing 1(1), 10–13 (2002)
14. Hauptmann, A.G., Gao, J., Yan, R., Qi, Y., Yand, J., Watctlar, H.D.: Automated Analysys of Nursing Home Observations. IEEE Pervasive Computing 3(2), 15–21 (2004)
15. Sixsmith, A., Johnson, N.: A Smart Sensor to Detect the Falls of the Elderly. IEEE Pervasive Computing 3(2), 42–47 (2004)
16. Ito, M., Katagiri, Y., Ishikawa, M., Tokuda, H.: Airy Notes: An Experiment of Microclimate Monitoring in Shinjuku Gyoen Garden. In: INSS, Networked Sensing Systems (2007)
17. Iwai, M., Mori, M., Tokuda, H.: Live! Commerce System: A marketing WSN enabling analyzing customers' attention in the real shops. In: INSS, Networked Sensing Systems (2008)
18. Sheikh, K., Wegdam, M., Sinderen, M.: Middleware Support for Quality of Context inn Pervasive Context-Aware System. In: Proceedings of Perware 2007 workshop (2007)
19. Korb, K., Nicholson, A.: Bayesian Artificial Intelligence. In: CRC Pr I Llc (2003)

A Framework for Quality of Context Management

Zied Abid, Sophie Chabridon, and Denis Conan

Institut TELECOM, CNRS UMR Samovar
9 rue Charles Fourier, 91011 Évry, France
Firstname.Lastname@institut-telecom.fr

Abstract. Context-aware computing has to deal with a huge amount of context data. Taking into account the quality of these data becomes a corner stone of an efficient context management solution. Information on the quality of context helps taking appropriate decisions and allows to identify uncertain context information saving processing time for deriving a pertinent description of the observed phenomenon.

This paper presents a work in progress for integrating Quality of Context in COSMOS (COntext entitieS coMpositiOn and Sharing) [4,13], a component-based framework for managing context data in ubiquitous environments, and illustrates it throughout the example of the composition of context information to implement a *network connectivity vs energy* adaptation situation.

Keywords: context-aware computing, quality of context, component-based middleware.

1 Introduction

With the proliferation of wireless connectable devices, the environment can be enriched with sensors acquiring a huge amount of context data that is to be analysed by computing systems. We consider context as being "any information that can be used to characterize the situation of entities (*i.e.* whether a person, place or object) that are considered relevant to the interaction between a user and an application, including the user and the application themselves"[7]. Context-aware computing allows to detect specific conditions requiring some adaptation actions. This calls for context integration and context abstraction methods. Context integration concerns the extraction of the most accurate context from a number of noisy and conflicting contexts. Context abstraction, or context reasoning, allows to derive a higher-level application-relevant context from a number of lower-level context data [15]. Taking into account the quality of context data becomes a corner stone of an efficient context management solution and has given rise to a large number of research works over the past decade. The importance of the quality of context as such for context-aware computing has first been raised by [3]. This concept has then been refined with a notion of worth: quality of context is any inherent information that describes context information and can be used to determine the worth of the information for a

K. Rothermel et al. (Eds.): QuaCon 2009, LNCS 5786, pp. 120–131, 2009.

specific application [10]. Information on the quality of context helps taking appropriate decisions and allows to identify uncertain context information saving processing time for deriving a pertinent description of the observed phenomenon.

This paper proposes to integrate Quality of Context in COSMOS[1] (*COntext entitieS coMpositiOn and Sharing*), which is our framework for managing context data in ubiquitous environments [4,13]. The COSMOS framework relies on theFRACTAL[2] component-based middleware [2]. FRACTAL presents some original features among which two are important to us: recursivity and component sharing. Recursivity (which has given its name to FRACTAL) allows components to be nested within composite components. With sharing, a given component instance can be included (or shared) by more than one component, saving memory and other system resources. COSMOS then provides the concepts of context node and context management policies translated into FRACTAL software components.

COSMOS reorganises the classical functionalities of a context manager to systematically introduce a 3-steps cycle of data collection, data interpretation, and situation identification. Although situation identification actions should not be too frequent, processing context information is an activity that must be conducted more often, while data gathering is a third activity that must be continuous. Thus, we have three different activities with different frequencies. We decouple as much as possible these activities in order to obtain a non-blocking and usable framework.

QoC is supported in COSMOS through the notion of QoC operator that can integrate various kinds of QoC parameter operators, dedicated to a particular QoC parameter *e.g.* Up-to-Dateness. This approach is generic and our framework can easily be extended by adding new operators. Moreover, we propose several modes to transmit QoC. The QoC data can be communicated either as metadata in a context report or separately. This contributes to provide a flexible framework that can be adapted to the requirements of various applications.

This paper is organised as follows. Section 2 presents the design of our COSMOS framework. Section 3 details the way we propose to integrate Quality of Context in COSMOS. Section 4 presents the case study of a *network connectivity vs energy* adaptation situation. Section 5 positions our work with respect to other research dealing with QoC in context management. Finally, Section 6 concludes this paper and identifies some perspectives.

2   Presentation of COSMOS

This section presents the foundations of our work by summarising the principles of the COSMOS framework. We present the basic building units for composing of a context policy, that is context nodes.

Concepts and properties of a context node. The basic structuring concept of COSMOS is the *context node* [4] which is a context information modelled by

---

[1] http://picoforge.int-evry.fr/projects/svn/cosmos/
[2] http://fractal.ow2.org

a software component. Context nodes are organised into hierarchies to form context management policies. Context nodes possess some properties which define their behaviour with respect to the context management policy. A context node can be *passive* or *active*. An active node is equipped with an activity to execute a given task. Communication into the hierarchy of context nodes may be bottom-up (*notification*) or top-down (*observation*). Observation (or notification) reports are messages formed of sub-messages and typed chunks. For instance, the information on the WIFI bit rate is stored in a chunk of type WifiBitRateChunk. A context node which receives data transmitted by a notification or an observation may be *blocking* or *non-blocking*. Non-blocking nodes propagate observations and notifications. Blocking nodes stop the traversal: for observations, the most up-to-date context information is transmitted without polling child nodes, and for notifications, context data is used to update the state of the node but parent nodes are not notified. COSMOS allows all kinds of combinations in the properties of context nodes (active/passive, observation/notification, blocking/non-blocking). This makes it possible to tune very precisely the level of computing resources used and to balance it with the requirements of applications.

COSMOS provides the developer with pre-defined generic context operators. They are organised following a typology: *Elementary operators* for collecting raw data, *memory operators*, such as averagers requiring a history of values or translation operators, *data mergers*, *abstract or inference operators*, such as additioners or thresholds operators. The only programming is in the context operators. Following the component-based software development principles, with a sufficiently large library of context operators, there should be no programming at all, but only declarative composition of context nodes.

Architecture of a context node. Each context node extends the abstract composite ContextNode depicted in Figure 1. The interfaces Pull and Push are the interfaces for the observation and the notification, respectively. The abstract composite ContextNode contains at least one operator (ContextOperator primitive component) as well as the message and activity managers. The Message

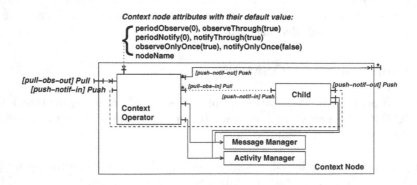

Fig. 1. Core architecture of a ContextNode component

Manager is in charge of handling the observation and notification reports which are sent and received by the component on the Pull and Push interfaces. The Activity Manager provides the support for dealing with active components. The Child (or children) is optional and can be a composite or primitive ContextOperator component. Child, Message Manager and Activity Manager components can be shared with other components, saving computing resources.

Pattern-oriented architecture of COSMOS. For mapping context policies to context node hierarchies, COSMOS follows well-known design patterns [8], *Factory method*, *Composite*, *Flyweight*, and *Singleton*, enabling a scalable, extensible and efficient architecture [13].

3   Quality of Context Management

We present in this section the way we propose to manage quality of context within the COSMOS framework. We detail the component based architecture that allows us to compute QoC parameters and to integrate them into the messages transmitted from context sources to the application.

3.1   Integrating QoC in COSMOS

In order to have a flexible framework, we propose three modes to transmit context information. The first two modes deal with QoC information while the last mode ignores it and allows to transmit context information without QoC. As shown in the case 1 of Figure 2, the first mode consists in injecting QoC information as meta-data into the context information itself before sending it to upper layers. This mode is useful to filter context according to a particular policy: for instance, no context message is sent if the *completeness* parameter is less than 50 %. The second mode sends QoC information independently from any context information in a separate message (cf. Figure 2-2). This mode enables to supervise the QoC of the system, with a limited overhead as only QoC data is computed and extracted. The third mode allows to transmit context information with standard child and/or parent components that cannot deal with QoC (cf. Figure 2-3). This mode is proposed to remove the cost of managing QoC information when this additional information is not necessary. It also ensures ascending compatibility with applications developed with previous versions of COSMOS not supporting QoC.

3.2   Architecture

We define a QoC ContextNode as a COSMOS composite ContextNode (cf. Figure 1) responsible for computing QoC parameter values. It is composed of Context Collectors which are themselves ContextNodes and a QoC Operator. A Context Collector collects raw meta-data coming from sensors or another part of the distributed system such as *Measurement Time, Source Location, Data accuracy* [12]. These raw meta-data are then transformed by the QoC Operator to deliver QoC parameters as shown below.

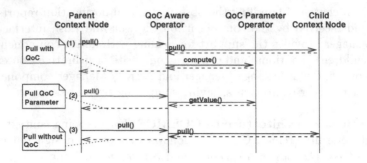

Fig. 2. Sequence diagram

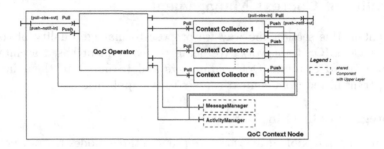

Fig. 3. QoC Context Node Architecture

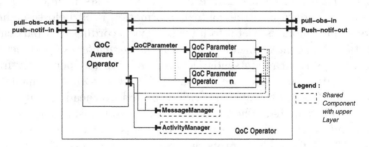

Fig. 4. QoC Operator Architecture

QoC Operator. The QoC Operator is responsible for extracting required data, computing QoC and supplying it to upper layers via the Message Manager (cf. Figure 3). As shown by the inner architecture of a QoC Operator (cf. Figure 4), raw meta-data coming from different Context Collectors get analysed by a QoC Aware Operator component which extracts relevant data and distributes them to QoC Parameter Operator components. Each QoC Parameter Operator computes a specific QoC parameter such as accuracy, precision, up-to-dateness, etc.

QoC Parameter Operator. The choice of the nature of the QoC Parameter Operator component depends on what type of QoC the application needs and

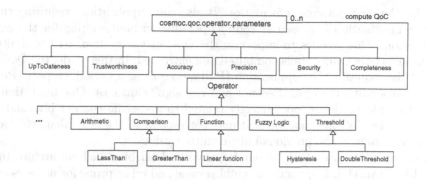

Fig. 5. Relation between QoC Parameters and COSMOS Operators

what computing methods are available. We propose in Figure 5 a first list of operators used in most of context-aware applications but other operators can easily be added. We give as an example a way to compute the Up-to-Dateness QoC parameter with function $\mathcal{U}(\mathcal{O})$ as presented in [12]:

$$\mathcal{U}(\mathcal{O}) = \begin{cases} 1 - \dfrac{Age(\mathcal{O})}{Lifetime(\mathcal{O})} & \text{if } Age(\mathcal{O}) < Lifetime(\mathcal{O}) \\ 0 & \text{Otherwise} \end{cases}$$

where $Age(\mathcal{O}) = tcurr - tmeasure(\mathcal{O})$, with $tcurr$ representing the current time and $tmeasure(\mathcal{O})$ the measurement time of object O.

After having been computed, QoC parameters are forwarded to a QoC Aware Operator which is responsible for sending QoC information to upper layers.

QoC Aware Operator. As introduced in Section 3.1, there are two modes to transmit QoC information that we now present in further details.

Adding QoC to context information. Raw QoC information is transmitted to QoC Parameter Operators to be processed (cf. Figure 2-1). Afterwards, once computed, QoC values can be either added in an existing message chunk or, taking advantage of the flexibility of COSMOS, they can be placed into a new message chunk dedicated to common QoC information like timestamp. In this mode, all context information messages are enriched with QoC meta-data. This mode is useful for applications interested in the QoC at the same time as the context information itself, that is when QoC information is systematically required. As a consequence, QoC parameters are strongly related to this context. This results in a more reliable and accurate analysis. However, this method requires time and resources to create a new message chunk or to update an existing one for each context information, which also increases the cost of the transmission of this information.

Sending QoC separately. Periodically, or on request, only the QoC is transmitted to upper layers (cf. Figure 2-2). This mode is well suited for applications that do not require QoC information with a strong timing constraint and that can wait for the next periodic information. It also suits applications requiring that context information is passed as soon as possible without waiting for the next notification. This enables to supervise the QoC of the system, with a limited overhead as only QoC data is computed and extracted. In this mode, the QoC is sent at the request of the application (Pull mode) or as a periodic report (Push mode), so sending the QoC does not add any significant cost. One limitation is that QoC information is not strongly related to a specific context information instance. The QoC information sent may correspond to the last calculated QoC or to an average of the previous untransmitted values.

Regarding the ease of use of our framework, describing such an architecture with FRACTALADL [11] would be cumbersome and error-prone for users. So we have defined a first version of a domain specific language (DSL) for describing the composition of context nodes and context processors [13] that we will extend with QoC declarations.

4   Application Scenario

In this section, we add QoC meta-data to an application scenario originally proposed for mobile commerce [13]. We consider a family shopping at a mall, each member of the family having a mobile device. This application allows them to share information, to consult product information, to download discount tickets, to be notified of advertisements, or to find the location of a product or a shop in the mall. The parents want their children to remain in the mall, with their devices connected as far as possible, so that everybody knows the location of the other family members. Nevertheless, a family member can disconnect for some periods of time in order to save their battery. The COSMOS context policy for this *network connectivity vs energy* scenario is shown in Figure 6. It involves different network technologies, such as Bluetooth or WIFI, and requires the application to adapt itself depending on network connectivity and context information availability. Each adaptation situation (in the upper part of Figure 6) is isolated in a context tree with the possibility of sharing sub-trees between policies.

We consider three QoC parameters in this scenario, *Trustworthiness*, *Up-to-Dateness* and *Precision* and detail below the way they are dealt with in our framework.

Trustworthiness. Adding trustworthiness to this system is essential to measure how much a device can trust data coming from a connection even before testing the link quality of the connection or starting to download data. We consider that trustworthiness depends on the identity of the data sender which may correspond to a family device, a mall information source, a shop in the mall, or be unknown. A trustworthiness manager observes the WIFI and Bluetooth managers that are directly connected to the system. The sender id can be extracted as a MAC[3], IP or DNS address (for WIFI sources), device name or BD_ADDR[4] (for Bluetooth sources).

---

[3] Media Access Control address.
[4] Bluetooth Device Address.

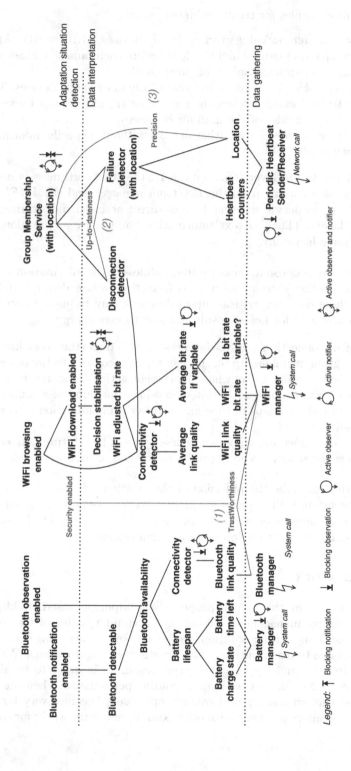

Fig. 6. Application scenario - Monitoring Network Connectivity and Battery Level

Here are some examples for trustworthiness values :

- 1 (100 %) *known device*: Value given to family devices. MAC or BD_ADDR address is unique and can be identified, therefore each family device can be registered into the application's configuration file.
- 0.5 (50 %) *verifiable device id*: Value given to mall or shops devices. These devices are trusted enough when their name or IP address can be verified (with the access provider of the mall for instance).
- 0 (0 %) *unknown device*: for all other devices without a specific information or not known at all.

In the data interpretation layer (cf. Figure 6-1), the trustworthiness is derived from the information coming from the *Bluetooth manager* and the *WIFI manager*. It then helps to decide whether to give direct access or not to incoming data to upper layers. Thus, context information has not to get up along the whole context node hierarchy.

Up-to-Dateness. We propose to measure the freshness of the information coming from the *Disconnection detector* as well as from the *Failure detector* ((cf. Figure 6-2). As these data are critical, up-to-dateness may be used to optimize connection management for better result and also to save energy.

- If failure or disconnection's up-to-dateness value is high (the event just happened), the group membership service can try to reconnect to the concerned sender/receiver in order to continue the current action (send/receive).
- If failure or disconnection's up-to-dateness value is medium, new connections can be postponed for a laps of time and another communication mechanism is to be used.
- If failure or disconnection's up-to-dateness value is low, future connections to the concerned sender/receiver can be postponed or even canceled.

Precision. In this example, the precision of the location information is used to adapt this information with appropriate display (cf. Figure 6-3). An application can have different map categories and precision degrees, and let the display manager choose the best map to promote location information.

5   Related Work

The work presented in this paper proposes a component-based middleware approach for context management taking into account Quality of Context as metadata. In the design of our solution, we have taken a particular care of the performance issue in terms of the cost of observations and notifications. We favor a flexible architecture and try to save system resources in order to be able to reach a good level of scalability without degrading performance when the number of observed context sources and context processors becomes very large. In this section, we compare our work with other middleware frameworks for context management.

The Context Toolkit is one of the first middleware framework for context management. It is based on event programming and widget concepts introduced by GUI (Graphical User Interfaces) [7]. In the same framework, all the following functionalities are grouped: The interpreter for composing and abstracting context information, the aggregator for the mediation with the application, the service for controlling application actions performed on the context, and the discoverer that acts as a registry. Following the same philosophy, interpretation and aggregation functionalities have to be programmed in monolithic blocks: One interpreter and one aggregator per application, independently of the number of widgets and the level of abstraction requested by the application. This implies a lack of flexibility, that can impact performance and scalability. Moreover, the management of system resources consumed by context management treatments and, in particular, activities management, is not addressed. Concerning the quality of context, the authors of the Context Toolkit have considered means to deal with the uncertainty of context data through its level of accuracy. Three complementary approaches were proposed: passing ambiguity on to applications; attempting to disambiguate context automatically; and attempting to disambiguate context manually. Only the latter manual approach has been further investigated.

MoCoA provides an environment for building context-aware applications for ad hoc networks based on sentient objects [14]. The low-level inference treatments are organised as data merging pipes. MoCoA only allows notifications, contrary to COSMOS that adds observations. The pipes are logically enclosed in sentients objects, including the control of system resources' consumption. But, contrary to COSMOS, MoCoA neither details nor provides any means to externally specify these controls. Pipe treatments are complemented with inference ones with facts and rules. Sensor fusion is then used to manage the uncertainty of data captured from the real-world and to derive higher-level context information. Fusion can perform a sum, an average function or might rely on a Bayesian network. However, the quality of context data is not considered as a first-class concept and remains restricted to a single certainty value.

Contextors [5] are software entities similar to data components, and their meta-data (describing the data quality) as well as their controllers (modifying the configuration) are available for both inputs and outputs. A Contextor is a Java class that is associated to an XML descriptor. Thus, the software framework builds, in an ad hoc manner, a container around the Contextor component. This ad hoc component model is implicit and not configurable (e.g. for managing system resources). For each Contextor using at least an activity, the local resource consumption can not be controlled. Furthermore, the sharing of context nodes supported by COSMOS is not addressed by Contextors. In addition, Contextors exchange control information in order to ask to stop or force the data notification for example. However, given that there is no explicit component model, it is impossible to introduce new configurations, such as some new attributes or control modes. In COSMOS, the structure and the life-cycle of components is finely managed by the FRACTAL controllers. One important aspect of the Contextor is the notion of data quality. Unlike to our solution, this quality meta-data can only be sent with the context data itself. We made the choice to provide different modes of transmission of QoC information for more flexibility and to allow to better tune the performance of the framework.

The list of quality parameters we consider is comparable to what is proposed in [12]. However, a specificity of our implementation framework is that it benefits from a component-based middleware following specific design patterns allowing to control very precisely resource consumption and performance.

6   Conclusion and Future Work

This paper presents a work in its early stage on Quality of Context management. We pay particular attention to performance and scalability issues by tailoring QoC management respectively to applications requirements and performance expectations. We are currently building a library of operators allowing to use combinations of rule-based and probabilistic solutions when appropriate in order to deal with uncertainty during the context abstraction process.

Regarding the ease of use of our framework, we have defined a first version of a domain specific language (DSL) for describing the composition of context nodes and context processors [13]. As future work, we intend to extend this DSL with QoC declarations.

Another research direction concerns the design of look-up mechanisms to find a child in the context hierarchy with a specific QoC level. Indeed, the component model we use is loosely typed, as mainly push and pull interfaces are defined. Therefore, our framework could benefit from a type system like Dream Types [1], allowing an operator to ask for a child node with a specific QoC type.

According to [6], specific probabilistic schemes are to be used at different abstraction levels. Depending on the amount of available knowledge, we envisage to experiment a method like FSI (Fuzzy Situation Inference) [9].

References

1. Bidinger, P., Leclercq, M., Quéma, V., Schmitt, A., Stefani, J.-B.: Dream Types: A Domain Specific Type System for Component-Based Message-Oriented Middleware. In: 4th ESEC/FSE Workshop on Specification and Verification of Component-Based Systems, Lisbon (Portugal) (September 2005)
2. Bruneton, É., Coupaye, T., Leclercq, M., Quéma, V., Stefani, J.-B.: The FRACTAL Component Model and Its Support in Java. Software—Practice and Experience, special issue on Experiences with Auto-adaptive and Reconfigurable Systems 36(11), 1257–1284 (2006)
3. Buchholz, T., Kupper, A., Schiffers, M.: Quality of context information: What it is and why we need it. In: 10th Int. Workshop of the HP OpenView University Association (HPOVUA). ACM, Geneva (2003)
4. Conan, D., Rouvoy, R., Seinturier, L.: Scalable Processing of Context Information with COSMOS. In: Indulska, J., Raymond, K. (eds.) DAIS 2007. LNCS, vol. 4531, pp. 210–224. Springer, Heidelberg (2007)

5. Coutaz, J., Rey, G.: Foundations for a Theory of Contextors. In: 4th International Conference on Computer-Aided Design of User Interfaces, Valenciennes (France), May 2002, pp. 13–34. Kluwer, Dordrecht (2002)
6. Dargie, W.: The Role of Probabilistic Schemes in Multisensor Context-Awareness. In: 5th IEEE Int. Conf. on Pervasive Computing and Communications. PerCom 2007, March 2007. IEEE Computer Society, Los Alamitos (2007)
7. Dey, A., Salber, D., Abowd, G.: A conceptual framework and a toolkit for supporting the rapid prototyping of context-aware applications. Special issue on context-aware computing in the Human-Computer Interaction Journal 16(2-4), 97–166 (2001)
8. Gamma, E., Helm, R., Johnson, R., Vlissides, J.: Design Patterns: Elements of Reusable Object-Oriented Software. Addison-Wesley, Reading (1994)
9. Haghighi, P.D., Krishnaswamy, S., Zaslavsky, A., Gaber, M.M.: Reasoning about context in uncertain pervasive computing environments. In: Roggen, D., Lombriser, C., Tröster, G., Kortuem, G., Havinga, P. (eds.) EuroSSC 2008. LNCS, vol. 5279, pp. 112–125. Springer, Heidelberg (2008)
10. Krause, M., Hochstatter, I.: Challenges in Modelling and Using Quality of Context (QoC). In: Magedanz, T., Karmouch, A., Pierre, S., Venieris, I.S. (eds.) MATA 2005. LNCS, vol. 3744, pp. 324–333. Springer, Heidelberg (2005)
11. Leclercq, M., Ozcan, A.E., Quema, V., Stefani, J.-B.: Supporting heterogeneous architecture descriptions in an extensible toolset. In: ICSE 2007: Proceedings of the 29th international conference on Software Engineering, pp. 209–219. IEEE Computer Society, Washington (2007)
12. Manzoor, A., Truong, H., Dustdar, S.: On the Evaluation of Quality of Context. In: Roggen, D., Lombriser, C., Tröster, G., Kortuem, G., Havinga, P. (eds.) EuroSSC 2008. LNCS, vol. 5279, pp. 140–153. Springer, Heidelberg (2008)
13. Rouvoy, R., Conan, D., Seinturier, L.: Software Architecture Patterns for a Context Processing Middleware Framework. IEEE Distributed Systems Online 9(6) (June 2008)
14. Senart, A., Cunningham, R., Bouroche, M., O'Connor, N., Reynolds, V., Cahill, V.: MoCoA: Customisable Middleware for Context-Aware Mobile Applications. In: Meersman, R., Tari, Z. (eds.) DOA 2006. LNCS, vol. 4276, pp. 1722–1738. Springer, Heidelberg (2006)
15. Ye, J., McKeever, S., Coyle, L., Neely, S., Dobson, S.: Resolving uncertainty in context integration and abstraction. In: ICPS 2008: 5th Int. Conf. on Pervasive Services, pp. 131–140. ACM, New York (2008)

An Abstract Processing Model for the Quality of Context Data

Matthias Grossmann, Nicola Hönle, Carlos Lübbe, and Harald Weinschrott

Universität Stuttgart, Institute of Parallel and Distributed Systems,
Universitätsstraße 38, 70569 Stuttgart, Germany
{grossmann,hoenle,luebbe,weinschrott}@ipvs.uni-stuttgart.de

Abstract. Data quality can be relevant to many applications. Especially applications coping with sensor data cannot take a single sensor value for granted. Because of technical and physical restrictions each sensor reading is associated with an uncertainty. To improve quality, an application can combine data values from different sensors or, more generally, data providers. But as different data providers may have diverse opinions about a certain real world phenomenon, another issue arises: inconsistency. When handling data from different data providers, the application needs to consider their trustworthiness. This naturally introduces a third aspect of quality: trust. In this paper we propose a novel processing model integrating the three aspects of quality: uncertainty, inconsistency and trust.

1 Introduction

Applications that process sensed data or integrate data from different independent data providers need to handle data with varying quality. To these applications it is crucial to measure data quality. Moreover, a measurement of quality can be beneficial for both applications and data providers. Applications can use it to exclude data that does not satisfy user needs and data providers could incorporate the quality of the provided data into their pricing policies. Often the quality of data hints at the costs of providing the data. It might be more expensive in terms of energy to provide an accurate, up-to-date data value of a sensor than an imprecise, possibly outdated value. Especially context-aware applications running on resource-limited mobile devices often have to trade quality against resource-consumption. These context-aware applications are the focus of the Nexus project [1].

A lot of research has been done on the subject of data quality. In most cases a metric of a certain quality aspect like uncertainty is used to define quality. In the context of the Nexus project, we have investigated three different aspects of quality: uncertainty, inconsistency and trust. In this paper we integrate all three aspects of quality into a single processing model.

The paper is organized as follows. Sect. 2 introduces the Nexus middleware and motivates the choice of the three quality aspects. Sect. 3 gives an overview of the related work. We define the three aspects of quality, namely certainty,

K. Rothermel et al. (Eds.): QuaCon 2009, LNCS 5786, pp. 132–143, 2009.

consistency and trust separately in Sect. 4 and we introduce operators used for formulating queries and an example scenario in Sect. 5. In Sect. 6, we explain the reasons for integrating the three quality aspects and present and evaluate a suitable processing model. Section 7 concludes with directions for further research.

2   The Nexus System

In the Nexus project [1], we provide a framework for managing global context models in an open platform, where a multitude of context data providers can integrate and share their context models. Due to the global characteristics and the high number of different context providers, our system is based on a distributed and scalable architecture.

We depict a simplified three-layer architecture of our system in Fig. 1. The bottom layer, i.e., the context information layer, consists of context data providers (CP) offering context information from various sources ranging from static information to sensor values. Thereby, different context providers may provide data with different levels of detail. In addition, this data can be based on different kinds of sensors. These two characteristics are the reason to specify the *uncertainty* of the data. Moreover, the fact that several context providers may provide data on the same phenomenon is the reason to specify the *inconsistency* of this data. Finally, in this open system, information about the *trust* in context providers is essential to estimate the value of the provided data. We explain the details of these three aspects of quality in Sect. 4.

The middle layer, i.e., the federation layer, is the platform for processing queries on the data provided by the context information layer. Thereby, the federation nodes (FN) on this layer provide the abstraction of a single data source to the applications (App) in the application layer. Processing of data quality is done based on the currently available data at the different providers. This processing is not influenced by limitations through network characteristics, e.g., interpolation mechanisms are incorporated to cope with high network delay.

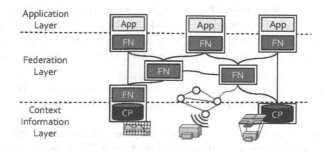

Fig. 1. Architecture of the Nexus system [1]

3   Related Work

Research has been done on quality of context data in various ways, however, distinguishing between different quality aspects is a novel concept. Papers typically talk about quality of data, meaning either certainty or consistency, and treating trust as a different issue. As we will outline in Sect. 4, our concept of certainty is mostly related to the area of sensor data and moving object databases (continuous domain, infinite number of alternatives) and consistency to the area of uncertain databases (discrete domain, finite number of alternatives). In the following, we present some works on quality of context data and show that our models, since they are based on these works, build on accepted knowledge.

Uncertainty is caused, e.g., by the characteristics of update protocols or trajectory simplification for position data or by the impreciseness of sensors. Probability density functions (PDFs) are commonly used to represent uncertain data. E.g., in case of GPS sensors[1], the PDF directly reflects the measuring accuracy. For uncertainty caused by update protocols, it is possible to specify a range and assume uniform distribution [2]. Although PDFs do not provide the most accurate means for representing uncertainty in all cases [3], we will use this well-known uncertainty model in this paper. Some papers in the area of uncertain data only consider simple range or nearest neighbor queries, e.g. [2,4], while others use more complex composed queries, e.g. [5,6]. From these two publications, we adopt the idea of composed queries and a one-dimensional (spatial) domain. In addition, for measuring uncertainty, we apply the well-known concept of differential entropy [7] as in [8].

In the area of inconsistent data, i.e., scenarios with a finite number of alternatives for a value, a concept sometimes called the possible worlds model is frequently used, e.g. in [9,10,11,12]. The advantage of the possible worlds model is that it formally can be used on top of the relational algebra without modifying the operators. Conceptually, for each possible combination of alternative values, a separate instance of the database (a possible world) is created and the query (consisting of standard relational algebra operators) is evaluated on each instance resulting in a set of alternative results.[2] We also rely on this model, but extend it by allowing uncertain attributes with PDFs. Cheng et al. also use the possible world model together with PDFs [8], but only regarded queries that can be classified either as returning a single value represented by a PDF or as returning entity sets, represented by possible worlds, so no actual integration of PDFs in the possible worlds model is required. For measuring inconsistency, a lot of related works [13,14] define distance-based metrics, which is also the idea of our model.

For modeling trust in context data providers, we use a simplified variant of the model from [15], which is based on Jøsang's opinion triangle [16].

---

[1] http://telecom.tlab.ch/~zogg/Dateien/GPS_Compendium(GPS-X-02007).pdf

[2] n values with alternatives in a database result in $O(2^n)$ possible worlds, so this model can be used to define the semantics of query processing, but in general not to actually implement it.

To the best of our knowledge, there is no work that tries to define a generic reference model for processing quality of context data, combining the different quality aspects and providing a single expressive interface for applications to quality of data.

4   Three Aspects of Quality of Data

In the following discussion, $o, o_1, o_2, \ldots$ denote objects. Objects are sets of attributes. The P attribute of o_1 is denoted by $o_1.P$. Here, we only regard attributes (called P) with scalar values, representing not only, e.g., temperature or other sensor measurement values, but also – as in the following discussion – position values in a one dimensional space. This is primarily for simplicity, but, depending on the data model, can also practically be used, e.g., for representing positions of cars on a highway [5].

Different data providers can manage the same object. We call the data providers $1, 2, \ldots$, and $o_1^2.P$ denotes the position of object o_1 according to provider 2.

The following definitions are chosen such that greater values correspond to an increase of the named quality aspect, i.e., greater values mean more *un*certainty, more *in*consistency and more trust.

4.1   Uncertainty

Imprecise sensors like GPS are the reason for uncertainty. So sensor values in general are not given exactly, but through a range of values. How this range is given depends on the sensor. In the following we assume that a probability density function (PDF) is given, but in the future, we plan to integrate more flexible representations.

We assume that data providers specify a normal PDF, however, due to the way we handle data of not fully trusted providers when fusing the quality aspects (cf. Sect. 6), we want to be able to express that, with some probability, we are not sure or do not know the value.

Definition 1. *An uncertain position P is represented by a special PDF $p : \mathbb{R} \to \mathbb{R}_0^+$ with $0 \leq \int_{-\infty}^{\infty} p(x)\, dx \leq 1$. With the probability $1 - \int_{-\infty}^{\infty} p(x)\, dx$, the value is unknown (NULL).*

Besides representing uncertain positions, we also require a means for measuring how uncertain a position is. For this, we adopt the concept of differential entropy from [7], which was already used for measuring quality of data in [8]. To be able to use this definition, we restrict the position PDF to have a lower bound l and upper bound u, with

$$p(x) \begin{cases} > 0, l \leq x \leq u \\ = 0, \text{otherwise} \end{cases}.$$

Definition 2. $u(P) = -\int_l^u p(x) \log_2 p(x)\, dx$ *is the uncertainty of position P.*

This definition restricts the form of the PDF and may not be adequate for cases where the probability for the value being NULL is greater than 0, however, as shown in Sect. 6, we only apply this definition to values directly retrieved from data providers, where these limitations are reasonable.

4.2 Inconsistency

Inconsistency occurs when different data providers offer the same datum, e.g., different sensors measure the same datum, or the buildings in a town are modelled from different organizations. This leads to a finite number of alternatives for one value. For measuring the inconsistency of two positions, we use the arithmetic mean of the smallest possible distance and the largest possible distance between the positions:

Definition 3. *The smallest and largest possible distance between two positions* P_1 *and* P_2 *are* $d_{\min} = \max(0, \max(l_2 - u_1, l_1 - u_2))$, $d_{\max} = \max(u_1, u_2) - \min(l_1, l_2)$. *The inconsistency of the two positions is*

$$i(P_1, P_2) = \frac{d_{\min} + d_{\max}}{2}.$$

4.3 Trust

We consider data providers to be differently reliable. The reliability of a data provider cannot be constituted globally, because it depends on the user and its preferences. In the Nexus project, we model trust as a triple (belief, disbelief, ignorance), where the three values are from the interval [0,1] and their sum is 1. In the following discussion we use a simplified version, where the disbelief value is always 0. In this case, it is sufficient to specify the belief value b (we trust the provider), the ignorance value (we cannot decide) is $1 - b$.

Definition 4. *The trust value of data provider* i *is given by* $b(i)$, $b : \mathbb{N} \rightarrow [0, 1]$.

5   Query Processing

As mentioned in Sect. 3, we use the possible worlds approach as basis for the query processing, but need to be able to represent uncertainty, so we have to extend the model to support an infinite number of possible worlds. This is subject to ongoing research, but for queries with simple selection predicates, the approach shown in Fig. 2 is reasonable. In addition to enumerate a finite number of possible worlds (boxes in Fig. 2), we allow uncertain attributes in a possible world (grey circles representing positions), so that a possible world in our model can represent an infinite number of exact possible worlds (shown in the bottom part of Fig. 2). In contrast to the original possible world model, we need to adapt operators for our approach. Fig. 2 shows a range query, which only a part of the o_1s represented by *PW1* fulfills, so the result of applying the query to *PW1* is an empty possible world (*PW3*), and a possible world with a modified position for o_1 (*PW2*).

The Nexus system is not only able to simply retrieve objects, but can also process more complex queries. It provides a set of generic operators, which is

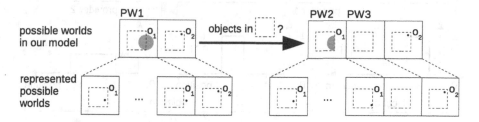

Fig. 2. Extending the possible worlds model to support uncertainty

similar to the relational algebra. The precise definition of the complete set is beyond the scope of this paper, but we briefly describe the operators used in the example scenario. Note that these operators only handle uncertainty, we explain in Sect. 6, why this is sufficient.

Selection σ_{pred}**:** The selection operator is equivalent to the selection operator of the relational algebra. It takes a list of objects as input and outputs a list containing all objects from the input list, which fulfill the predicate *pred*. When applied to uncertain data, objects fulfill the predicate with some probability, and objects are included in the result list with this probability, i.e., σ can create several alternative results (possible worlds) and is an entity-based non-aggregate operator according to the classification in [8]. As previously explained, it may be necessary to modify uncertain attribute values. NULL values are handled as in SQL: When *pred* evaluates to *unknown*, the object is not included in the result.

Sorting $sort_{expr}$**:** The sorting operator sorts a list of objects. *expr* is an expression based on attributes of an object. It is evaluated for each object in turn, and the objects are sorted according to the results. Like the selection operator, sorting can create several alternative results when applied to uncertain data. The probability of a result list is determined by the probability that evaluating *expr* in sequence on all objects of this list results in a sorted list. Sorting is an entity-based aggregate operator.

Fetch $fetch_n$**:** The fetch operator just cuts a list of objects to the first n objects. It does not evaluate attributes like the other two operators do, thus does not require an adaption to handle uncertain data. We use the fetch operator in conjunction with sorting to implement a nearest neighbor query.

5.1   Example Scenario

Fig. 3 shows the example scenario. Two providers 1 and 2 store two objects o_1 and o_2. For o_2, each provider offers a representation, these two representations are different.

The uncertainties of the positions are $u(o_1^1.P) = u(o_2^1.P) = 0, u(o_2^2.P) = 1$, the inconsistency of $o_2.P$ is $i(o_2^1.P, o_2^2.P) = 1$. We want to answer the query, which of the objects located between the positions 1 and 3 is closest to position 0, more formally

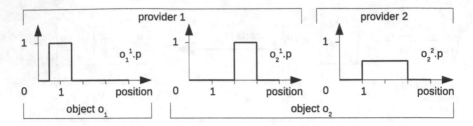

Fig. 3. Example scenario: PDFs of the positions of o_1 and o_2

$$fetch_1(sort_{dist(0, o.P)}(\sigma_{1 \leq o.P \leq 3}[o_1, o_2])) \ .$$

dist calculates the distance between its arguments. As in this scenario, the first argument is the position 0, the result has the same PDF as the second argument.

As explained above, it may be necessary to adapt the PDF for the position during selection. For a selection of the form $\sigma_{l \leq P \leq u}$, we do this the by narrowing the range, where the PDF is > 0, to the interval $[l, u]$ and multiplying the resulting function with a constant factor, so that the integral equals to 1:

$$p'(x) = \begin{cases} \frac{p(x)}{\int_l^u p(x)\, dx}, & l \leq x \leq u \\ 0, & \text{otherwise} \end{cases}$$

For evaluating *sort*, we must calculate the probability that a distance D_2 is greater than an other distance D_1. When D_1 and D_2 are represented by two PDFs d_1 and d_2, the probability for $D_2 > D_1$ is[3]

$$\int_{-\infty}^{\infty} \int_{x_1}^{\infty} d_1(x_1) d_2(x_2)\, dx_2\, dx_1 \ . \tag{1}$$

In the given scenario, we cannot be sure if o_1 actually fulfills the selection predicate, and – according to the data of provider 2 – there is a chance that o_2 is closer to 0 than o_1. Obviously, the probability for o_1 to be closer to 0 is much higher than for o_2. However, the probability for o_1 to fulfill the selection predicate is only 0.5, so we expect the probability for o_2 being the result of the query to be only a little bit above 0.5.

6   Processing Model

In Sect. 4, we presented approaches for representing and measuring uncertainty, inconsistency and trust on the data provider level. To be able to process complex queries like the one presented in the previous section, we need to address two additional questions: how to account for the quality aspects during the processing of queries and how to measure the quality of the final result set.

---

[3] $d_1(x_1) d_2(x_2)$ is the combined PDF for D_1 and D_2. To derive the probability for $D_2 > D_1$, we need to integrate over the area where $x_2 > x_1$.

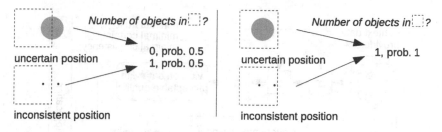

Fig. 4. Measuring the quality of the query result

The straightforward attempt to solve the first problem would be to define separately for each operator, how each quality aspect is handled. When, e.g., the selection operator is applied to an uncertain attribute, the uncertainty selection operator would be invoked, and for an inconsistent attribute the inconsistency selection operator. However, this approach cannot handle information that is both uncertain and inconsistent, like $o_2.P$ in Fig. 3. Therefore, we need a more integrated concept, which can deal with all three quality aspects simultaneously.

To measure the quality of result sets, in some cases, it is possible to directly apply the definitions to query results. When an object with an uncertain position is present in a result set, the uncertainty model presented in Sect. 4.1 can be used to represent its position. Likewise, the inconsistency model from Sect. 4.2 can be used when two different values for the same attribute of an object are in the result set. However, when only one value qualifies for the result set, the inconsistency information gets lost. To use the trust model from Sect. 4.3, the trust value for the provider has to be assigned to each attribute he provides to the result. However, in more complex situations, these definitions are not suitable. Figure 4 shows on the left hand side a situation where we are not sure if the answer to the query *How many objects are located inside the dashed square?* is 0 or 1. This should somehow be reflected by the result's quality, but is unclear if this is uncertainty or inconsistency, because exactly the same result can be caused by uncertainty (top) or by inconsistency (bottom). On the right hand side, we have the same situation with a slightly shifted square for the query. In this case, we can be sure that the result of the query is 1, so the quality of the result should be optimal, although the data used for answering the query is uncertain or inconsistent.

To address these two problems, we are investigating the approach depicted in Fig. 5. The main idea is to combine the three aspects before the actual query processing takes place, and define query processing and the result's quality based on the possible worlds model. In the following, we discuss reasons for choosing this approach.

Viewed from the perspective of query processing, uncertainty and inconsistency describe similar phenomena – there exist several alternatives for one value. In the case of uncertainty, the number of alternatives is possibly infinite, whereas in the case of inconsistency, a finite number of alternatives exist. In that sense, uncertainty is a generalization of inconsistency and both can be expressed by an uncertainty model.

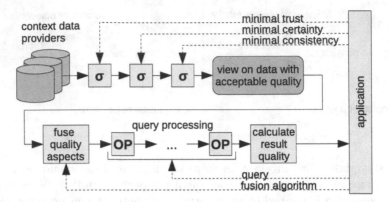

Fig. 5. Processing model

When expressing inconsistency as uncertainty, we basically add all PDFs for an attribute. Thereby we have to weight the individual PDFs of the data providers, in the most simple case with the reciprocal of the number of data providers. In our case, however, we can refine the weighting using the trust values, so that PDFs from trustworthy data providers gain a higher weight than those from lesser trusted ones. This meets the supposable expectation of users that information of trustworthy data providers is more likely to be true.

In more detail, the approach consists of the following steps:

1. Applications or users may want to specify minimum requirements for certainty, consistency and trust for the data used for processing the query. Three additional selections are performed before the actual query is processed which result in a subset of the original data set, that fulfills the quality constraints. Note that the selection of sufficiently trusted data providers has to be done before evaluating the consistency constraint, otherwise, untrusted providers would be able to force the removal of attributes from the subset by providing incorrect representations of the attribute, thus decreasing the consistency.

2. The three quality aspects are combined based on the uncertainty model. When the providers $1, \ldots, n$ provide values for the position of an object o, the resulting position is

$$o.p(x) = \frac{1}{n} \sum_{i=1}^{n} b(i)(o^i.p(x)) \ .$$

Inconsistency is incorporated by averaging the representations, trust by weighting them. Note that $\int_{-\infty}^{\infty} o.p(x)\, dx$ may be smaller than 1 (cf. Sect. 4.1). In some cases, applications may require a different fusion algorithm, so we provide a way for the application to specify the algorithm to use.

3. The calculation of the quality of the query's result is still an open issue, but using some extension of an entropy based approach seems to be promising.

It is not necessary to use an additional selection here, the application itself can decide whether the quality of the result is sufficient or not and discard the result in the latter case.

An additional benefit of this approach is that the combined data quality model is closely related to models typically used in the literature, which allows us to define the semantics of our operators based on well understood concepts.

6.1   Revisiting the Example Scenario

In this section, we describe how the query in Sect. 5.1 is processed using our processing model. We use the notation $[o_1, \ldots, o_n]_p$ for a result list generated with probability p.

For the first example, we trust each data provide fully, i.e., $b(1) = b(2) = 1$ and we do not use restrictions on certainty, consistency and trust. Thus, fusing the data of the two providers results in $o_1 = o_1^1$ and o_2 with

$$o_2.p(x) = \begin{cases} 0.25, 1 \leq x < 2 \\ 0.75, 2 \leq x \leq 3 \\ 0, \quad \text{otherwise} \end{cases} .$$

Fig. 6 shows the intermediate results after each operator of the query and the final result. σ does not modify o_2, because its position lies completely inside the requested area, o_1, however, becomes o_1' with

$$o_1'.p(x) = \begin{cases} 2, 1 \leq x \leq 1.5 \\ 0, \text{otherwise} \end{cases} .$$

o_1' is closer to 0 than o_2 with a probability of $\frac{15}{16}$ according to (1). The probability of o_2 being the final result of the query is a little bit higher than 0.5 as expected in Sect. 5.1.

For the second example, shown in Fig. 7, we use the trust values $b(1) = 0.5$ and $b(2) = 1$. This results in the following situation after fusing the data:

$$o_1.p(x) = \begin{cases} 0.5, 0.5 \leq x \leq 1.5 \\ 0, \quad \text{otherwise} \end{cases} \qquad o_2.p(x) = \begin{cases} 0.25, 1 \leq x < 2 \\ 0.5, \quad 2 \leq x \leq 3 \\ 0, \quad \text{otherwise} \end{cases} .$$

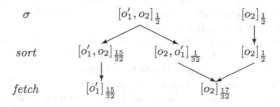

Fig. 6. Processing the query $(b(1) = b(2) = 1)$

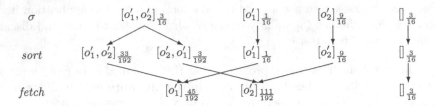

Fig. 7. Processing the query $(b(1) = 0.5, b(2) = 1)$

Because we do not fully trust provider 1, $o_1.P$ is NULL with probability 0.5 and $o_2.P$ with probability 0.25. o_1 fulfills the selection predicate with a probability of 0.25, o_2 with a probability of 0.75, so the selection also modifies $o_2.p$:

$$o_1'.p(x) = \begin{cases} 2, 1 \leq x \leq 1.5 \\ 0, \text{otherwise} \end{cases} \qquad o_2'.p(x) = \begin{cases} \frac{1}{3}, 1 \leq x < 2 \\ \frac{2}{3}, 2 \leq x \leq 3 \\ 0, \text{otherwise} \end{cases}.$$

Equation (1) yields $\frac{11}{12}$ for the probability of o_1' being closer to 0 than o_2'.

7 Conclusions and Future Work

In this paper we presented an abstract processing model for the quality of context data. We explained the three quality aspects uncertainty, inconsistency, and trust we use in the Nexus project and showed how the possible worlds model can be applied in this scenario.

In the future we plan to integrate more advanced models for the different quality aspects and multidimensional coordinates. The scheme for computing the quality of the overall result is also subject to ongoing research.

Acknowledgments. The work described in this paper was partially supported by the German Research Foundation (DFG) within the Collaborative Research Center (SFB) 627.

References

1. Lange, R., Cipriani, N., Geiger, L., Großmann, M., Weinschrott, H., Brodt, A., Wieland, M., Rizou, S., Rothermel, K.: Making the world wide space happen: New challenges for the Nexus context platform. In: PerCom Workshops, IEEE Computer Society, Los Alamitos (to appear, 2009)
2. Pfoser, D., Jensen, C.S.: Capturing the uncertainty of moving-object representations. In: Güting, R.H., Papadias, D., Lochovsky, F.H. (eds.) SSD 1999. LNCS, vol. 1651, pp. 111–132. Springer, Heidelberg (1999)
3. Lange, R., Weinschrott, H., Geiger, L., Blessing, A., Dürr, F., Rothermel, K., Schütze, H.: On a generic inaccuracy model for position information. In: Rothermel, K., Fritsch, D., Blochinger, W., Dürr, F. (eds.) QuaCon 2009. LNCS, vol. 5786, pp. 76–87. Springer, Heidelberg (2009)

4. Cheng, R., Prabhakar, S., Kalashnikov, D.V.: Querying imprecise data in moving object environments. In: Dayal, U., Ramamritham, K., Vijayaraman, T.M. (eds.) ICDE, pp. 723–725. IEEE Computer Society, Los Alamitos (2003)
5. de Almeida, V.T., Güting, R.H.: Supporting uncertainty in moving objects in network databases. In: Shahabi, C., Boucelma, O. (eds.) GIS, pp. 31–40. ACM, New York (2005)
6. Wolfson, O., Sistla, A.P., Chamberlain, S., Yesha, Y.: Updating and querying databases that track mobile units. Distributed and Parallel Databases 7(3), 257–387 (1999)
7. Shannon, C.E., Weaver, W.: The mathematical theory of communication, 4th edn. The University of Illinois Press (1969)
8. Cheng, R., Kalashnikov, D.V., Prabhakar, S.: Evaluating probabilistic queries over imprecise data. In: Halevy, A.Y., Ives, Z.G., Doan, A. (eds.) SIGMOD Conference, pp. 551–562. ACM, New York (2003)
9. Barbará, D., Garcia-Molina, H., Porter, D.: The management of probabilistic data. IEEE Trans. Knowl. Data Eng. 4(5), 487–502 (1992)
10. Agrawal, P., Benjelloun, O., Sarma, A.D., Hayworth, C., Nabar, S.U., Sugihara, T., Widom, J.: Trio: A system for data, uncertainty, and lineage. In: Dayal, U., Whang, K.Y., Lomet, D.B., Alonso, G., Lohman, G.M., Kersten, M.L., Cha, S.K., Kim, Y.K. (eds.) VLDB, pp. 1151–1154. ACM, New York (2006)
11. Benjelloun, O., Sarma, A.D., Halevy, A.Y., Theobald, M., Widom, J.: Databases with uncertainty and lineage. VLDB J 17(2), 243–264 (2008)
12. Cheng, R., Chen, J., Xie, X.: Cleaning uncertain data with quality guarantees. PVLDB 1(1), 722–735 (2008)
13. Yu, H., Vahdat, A.: Design and evaluation of a conit-based continuous consistency model for replicated services. ACM Trans. Comput. Syst. 20(3), 239–282 (2002)
14. Xu, B., Wolfson, O., Chamberlain, S.: Spatially distributed databases on sensors. In: GIS 2000: Proceedings of the 8th ACM international symposium on Advances in geographic information systems, pp. 153–160. ACM Press, New York (2000)
15. Gutscher, A., Heesen, J., Siemoneit, O.: Possibilities and limitations of modeling trust and reputation. In: Möller, M., Roth-Berghofer, T., Neuser, W. (eds.) WSPI, CEUR Workshop Proceedings, vol. 332 (2008), CEUR-WS.org
16. Jøsang, A.: Artificial reasoning with subjective logic. In: Proceedings of the Second Australian Workshop on Commonsense Reasoning (1997)

Using Quality of Context to Resolve Conflicts in Context-Aware Systems*

Atif Manzoor, Hong-Linh Truong, and Schahram Dustdar

Distributed Systems Group, Vienna University of Technology
{manzoor,truong,dustdar}@infosys.tuwien.ac.at

Abstract. Context-aware systems in mobile and pervasive environments face many conflicting situations while collecting sensor data, processing sensor data to extract consistent and coherent high level context information, and disseminating that context information to assist in making decisions to adapt to the continuously evolving situations without diverting human attention. These conflicting situations pose stern challenges to the design and development of context-aware systems by making it extremely complicated and error-prone. Quality of Context parameters can be used to cope with these challenges. In this paper, we discuss the conflicting situations that a context-aware system may face at different layers of its conceptual design and present the conflict resolving policies that are defined on the basis of the Quality of Context parameters. We also illustrate how these policies can be used in different conflicting situations to improve the performance and effectiveness of context-aware systems.

1 Introduction

The vision of pervasive environments is characterized with a plethora of computation and communication enabled sensing devices that are embedded in our daily-life objects. The main objective of these devices is to facilitate the work of user by acting as a smart assistant for them. Many research efforts have been undertaken to fulfill these requirements and context management system frameworks are divided in different conceptual layers. These conceptual layers are assigned the task of collecting raw sensor data, extracting high level context information from this data, aggregating and storing context information after eliminating redundant and inconsistent context, providing this information to interested applications and users, and finally applications and users take actions to adapt themselves to this information as described in [1].

Different conflicting situations can arise during the execution of the aforementioned tasks. These conflicting situations strongly affect the capability of context-aware systems to adapt to the evolving situation in pervasive environments. Earlier systems have used some simple strategies such as drop all, drop last, drop first [24], involved user to resolve conflicts [7], or did the mediation on the basis of some predefined static policies [17]. These strategies may slow down the process of decision making, distract users, or discard some important context objects as well. Moreover, context conflicts

---

* This research is partially supported by the European Union through the FP6-2005-IST-5-034749 project WORKPAD.

K. Rothermel et al. (Eds.): QuaCon 2009, LNCS 5786, pp. 144–155, 2009.

Table 1. Layers of the conceptual framework of a context management system and conflicts that can arise on those layers

Conceptual framework layers [1]	Conflicts	Examples
Context Acquisition	Conflicts in making selection among different sensors using different techniques to collect context [6, 5]	GPS and GSM collect location information of a mobile user with different level of accuracy
Processing	Conflicts in extracting high level context information [22, 21, 8]	Sensor data shows that someone is present in two different locations
Context Distribution	Conflicts in context information aggregation [18]	Redundant and inconsistent data reaching a node from different routs
Application	Conflicting interests of applications [17]	Different preferences set by two users present in living room

cannot be resolved at design time [4] and need a strategy that can dynamically handle them at runtime without distracting users.

Quality of context (QoC) is defined as "any information that describes the quality of information that is used as context information" [3]. QoC can be used to devise the policies to resolve the conflicts at different layers of a conceptual framework of context-aware systems as shown in Table 1. Later QoC is also defined as "any inherent information that describes context information and can be used to determine the worth of information for a specific application" [13]. In [14], we have classified QoC in QoC parameters and QoC sources. QoC parameters, such as up-to-dateness, trust-worthiness, completeness, and significance, are used to indicate the quality of context information. QoC sources like source location, measurement time, source state, and source category are used to evaluate those QoC parameters. In this paper we analyze generic conflicting situations that can occur at different layers of a context-aware system and propose the conflict resolving policies based on the quality of context parameters. We also present how these conflict resolving policies can be used and describe the prototype implementation of our system that have used these policies. We have performed the experiments to evaluate these policies. We observed those policies that used the combination of different QoC parameters considering the perspective of the use of context information are more effective.

The rest of the paper is organized as follows: Section 2 discusses the conflicts that a context management system can face at different layers of its conceptual architecture. Section 3 presents the policies that we have defined to resolve the conflicts discussed in Section 2. The implementation detail of our Quality-Aware Context Management Framework is discussed in Section 4. We have presented the experiments and evaluation of our system in Section 5. Section 6 gives an overview of related work and compares them with our approach. Finally, we conclude the paper and discuss our future work in Section 7.

2   Conflicting Situations in Context Management Systems

In this section we discuss the conflicts that can take place at different layers of context management system as presented in [1] and how these conflicts can affect the performance of context-aware applications. Table 1 provides the summery of these conflicts.

2.1   Context Acquisition

In pervasive environments the volume of data generated by sensors makes the analysis of context impossible for a human [6]. Sensor data may also differ with each other considering the frequency of updating context, the capability of a sensor to collect the context of an entity, the accuracy of a method that is used by sensors, representation format, and the price of context information [6, 5]. For example, location information of a mobile user can be gathered using GPS and GSM methods. Problems also arise due to the mobility of sensors along with entities in pervasive environments. We cannot permanently rank a sensor to collect the context of a particular entity. So there is a need for a strategy that can dynamically decide which sensor is more reliable to collect the context of a certain entity at some specific time. QoC parameters that have been dynamically evaluated from the information about the source of context can be used to resolve the conflicts in such situation.

2.2   Processing

In the processing layer, high level context is extracted from low level sensor data. Sensor data cannot be presented directly to applications. It needs to be filtered, fused, correlated, and translated to extract the higher level context data and detect the emergent events [22]. Some works, such as presented in [21], have made the supposition that sensors along with the produced data can also be used to estimate its reliability and self-confidence. These metrics are used to do the reasoning to extract high level context information. In [8], the advertised probability of correctness of context sources is used to do the reasoning to extract a single piece of context information by combining the information from different providers. QoC parameters that provide information about up-to-dateness, trustworthiness, significance and completeness can replace those metrics and make the reasoning on data more meaningful and realistic to resolve conflicts.

2.3   Context Distribution

The high mobility of sensors, unreliable wireless connections, and the nature of tasks in pervasive environments result in the acquisition of a lot of redundant and conflicting context. This redundant and conflicting context not only results in the wastage of scare resources but also can lead to undesired behavior of context-aware applications. Simple conflict resolving policies, such as drop first, drop all, can result in deleting some valuable information. In critical situation, such as a context-aware ubiquitous home for patients [12] and telehealth applications [11], loss of information can result in severe situations for the people using it. Decision can better be made to discard or keep a context object on the basis of policies defined using these QoC parameters.

2.4   Application

Context-aware applications use context information to adapt their behavior to user needs and changes in the environment. If conflicts are not resolved in context information at the earlier stages, the applications that take actions on the basis of that context information get in conflict while making decisions. Context-aware applications can also get in conflicts due to different priorities set by users. Different strategies are used to resolve their behavior as presented in [17]. Information about the up-to-dateness, trustworthiness, completeness, and significance of context information make it easy to resolve conflicts and make decisions on the basis of that context information.

3   QoC Based Conflict Resolving Policies

The main consideration of these policies is to resolve the conflicts in such a way that the decision should have been taken in the favor of context objects that contain the context information of the highest quality. This quality of context information is characterized by QoC parameters. Table 2 shows the equations that have been used to evaluate these QoC parameters in range [0..1] as described in [14]. If the user of these policies wants more than one context object that have quality higher than a specific value then he can specify the threshold value in range [0..1] and all the context objects that have quality higher than that threshold value are selected. We have also taken into account the user centered design of context-aware systems and tried that human users of the system should not be distracted during the execution of these policies. In this section we discuss the fundamental policies based on different QoC parameters.

Table 2. Some of the QoC parameters defined and evaluated in [14]

QoC Parameters	Equations for the evaluation of QoC Parameters for context object O
Up-to-dateness	$\begin{cases} 1 - \frac{Age(\mathcal{O})}{Lifetime(\mathcal{O})} & : \quad if \ Age(\mathcal{O}) < Lifetime(\mathcal{O}) \\ 0 & : \quad otherwise \end{cases}$
Trustworthiness	$\begin{cases} (1 - \frac{d(\mathcal{S},\mathcal{E})}{d_{max}}) * \delta & : \quad if \ d(\mathcal{S},\mathcal{E}) < d_{max} \\ undefined & : \quad otherwise \end{cases}$
Completeness	$\begin{cases} \frac{\sum_{j=0}^{m} w_j(\mathcal{O})}{\sum_{i=0}^{n} w_i(\mathcal{O})} & : \quad if \ m \ and \ n \ are \ finite \\ 0 & : \quad otherwise \end{cases}$
Significance	$\dfrac{CV(\mathcal{O})}{CV_{max}(\mathcal{O})}$

3.1 Up-to-Dateness Based Policy

Up-to-dateness indicates the degree of rationalism to use a context object at a specific instance of time. We have calculated up-to-dateness of a context object as the ratio between the age of that context object and the lifetime of the type of context information contained by that context object. This metric can be useful for resolving conflicts in the context objects that change their values very rapidly, e.g., the location of a fast moving vehicle. In this case, it will be more suitable to use the context object with the highest value of up-to-dateness. Whereas, up-to-dateness will not have a significant role in the case of conflicts in static information that have been profiled in the system, e.g., information about the structure of a smart home.

3.2 Trustworthiness Based Policy

Trustworthiness is the degree of the suitability of a sensor to collect the context of a specific type. We have calculated the trustworthiness of a context object on the concept of space resolution and accuracy of sensor to measure that type of information. This concept is particularly useful in resolving the conflict when we have more than one sensor collecting the context of same entity or event. For example, we have temperature sensors at different places in the living room of a smart home that is built to provide comfortable life to old people. The sensors that are installed near the electric radiator heater will be sending the higher value of the temperature of living room as compared to the sensors in the other places in the living room. To provide a comfortable temperature in the room we will be more relying on the readings of the sensors that are closer to the sitting area than the sensors in the far off corners of the living room and sensor near the radiator.

3.3 Completeness Based Policy

The completeness of context information indicates that all the aspects of context information have been presented by a context object. We have evaluated the completeness of a context object as the ratio of the sum of the weights of available attributes of context object to the sum of the weights of the total number of attributes of the context object. Completeness of a context object is particularly important to get the complete picture of the current situation of the real world. According to this policy decision is made on the basis of that context object which has more complete information about the current situation.

3.4 Significance Based Policy

Significance measures the worth or preciousness of a context object. It is particularly important to mention this metric when there is a context object of high critical value. For example, if smoke sensors detect heavy smoke in the bedroom, it will be an information of high significance. This metric can be used to generate events that need prompt actions from the applications. Applications can specify that the context objects with high values of significance should be reported on a priority basis.

Apart from the above mentioned fundamental policies, policies can also be defined based on two or more QoC parameters depending on the requirements of a particular application. For example, a policy can also be defined by combining QoC parameters, such as up-to-dateness and trustworthiness. In such policies an average value of the mentioned QoC parameters is used to make decisions. For example, if a context aggregator uses a policy based on the combination of the up-to-dateness of a context object with the threshold value of 0.8, then all the context objects having an average of the value of up-to-dateness and the value of trustworthiness of more than 0.8 will be selected. Users of conflict resolving policies set threshold values according to their requirements considering the perspective of the use of context information.

4 Implementation

Figure 1 shows the components of our context management system corresponding to the conceptual framework layers and data flow among those components. Components used to evaluate and annotate QoC parameters and conflict resolving policies can be used with any system component. *QoC Evaluator* receives context objects as XML elements and evaluates QoC parameters for those context objects. QoC parameters are normalized to have values in range [0..1] and *Context Annotator* annotates context objects with those QoC parameters as we had presented in [14]. Figure 2 shows a context object annotated with QoC parameters. QoC parameters along with QoC based conflict resolving policies are used to resolve conflicts. Guidelines to select a policy in different conflict resolving situations are provided to the system as an input file. For example,

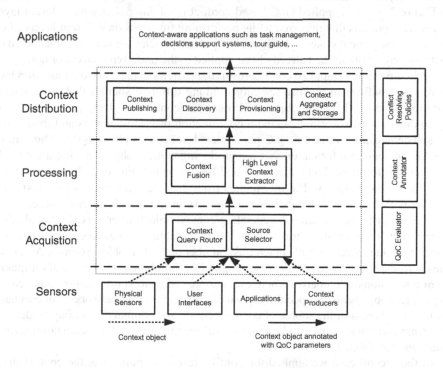

Fig. 1. Components of QCMF corresponding to different layers

it can be mentioned that the selection among different sources of a particular type of context information should have been done on the basis of the combination of trustworthiness and up-to-dateness based policies.

We developed our prototype as part of the implementation of the EU project WORK-PAD [20] based on COSINE [10]. Java2 ME (CDC 1.1 profile) was used for its development to make it able to run on mobile devices. Our context information model, to manage the context information in disaster response, was designed as XML schema. Context information is also stored as XML elements. MXQuery and KSOAP2, which have a low memory foot print to be executed on mobile devices, were used for processing XML data. In this prototype, we dealt with high level context information. The components dealing with low level context, such as context fusion and the high level context extractor, were not implemented yet.

5   Experiments and Evaluation

In our simulated environment, a team of five workers was performing rescue activities in response to a flood in a city. Those workers were randomly moving on the flood site and after every minute they were sending a context object to their team leader. This context object was of type infrastructure and contained information about the usability of a square in the city. Our context management system evaluated QoC parameters for that context object and annotated that context object with those QoC parameters as shown in Figure 2. We had applied QoC based conflict resolving policies on different layers to resolve conflicts that occurred while performing functions on different layers of our context management system. While applying those policies we have not considered the fact that those policies had already been applied to the underlying layers or not.

In the first case we applied conflict resolving policies at the context acquisition layer to resolve conflicts in making selections among different sources that were sending the aforementioned context objects. We used the conflict resolving policies based on up-to-dateness, trustworthiness, and a combination of up-to-dateness and trustworthiness. The threshold value for QoC-parameters has been specified as 0.9. Thus, all the sources of context information that are producing context objects having a value of a QoC parameter more than a threshold had been selected. Figure 3 shows the number of context objects received in 60 minutes from the selected sources of context information with increase in number of workers. As in our simulated environment every source of context information is generating context objects after a fixed interval of one minute. The number of context objects having a value of up-to-dateness more than a specified value increases with a increased number of context object sources. As a result the up-to-dateness-based conflict resolving policy did not seem to be useful for making source selections in scenarios where every source of context information was generating context objects after a specific interval of time, e.g., sensor networks. Policies based on trustworthiness of the source of context objects and combinations of up-to-dateness and trustworthiness proved to be more useful for the context data acquisition layer as mentioned in Table 3.

In the second case we applied the conflict resolving policies at the context distribution layer to make efficient use of the context stored by the context aggregator. The

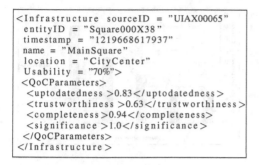

```
<Infrastructure  sourceID = "UIAX00065"
  entityID = "Square000X38"
  timestamp = "1219668617937"
  name = "MainSquare"
  location = "CityCenter"
  Usability = "70%">
  <QoCParameters>
    <uptodatedness>0.83</uptodatedness>
    <trustworthiness>0.63</trustworthiness>
    <completeness>0.94</completeness>
    <significance>1.0</significance>
  </QoCParameters>
</Infrastructure>
```

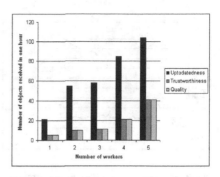

Fig. 2. XML representation of context object of type Infrastructure

Fig. 3. Source selection using different QoC-based conflict resolving policies

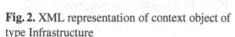

context aggregator received context objects from five workers that were sending data after random intervals of less than one minute. The context aggregator was initiating a cleaning service after every minute and conflicting context objects that did not meet the specified policy criteria were deleted. Conflict resolving policies based on up-to-dateness, trustworthiness, and quality, i.e., the combination of up-to-dateness and trustworthiness, have been used and the threshold value of 0.85 has been set for those policies. Figure 4 shows the number of context objects that were currently stored in the context store using the aforementioned policies and threshold value. As it is apparent from Figure 4, using the policy based on trustworthiness did not prove to be very useful as some context objects that have been captured long time ago still have higher values of trustworthiness and are uselessly kept in the context store. Using the up-to-dateness-based policy proved to be the same as keeping the latest context objects. It deleted the old context objects which can result in the loss of some important context information. Finally, we have used a quality policy based on the combination of both up-to-dateness and trustworthiness to detect useless context objects. With this policy we had not only been able to detect a higher number of useless context objects but also kept the context objects of high trustworthiness that are highly valued in making any decision on the basis of context information. From this observation we conclude that the conflict resolving policy based on the combination of more than one QoC parameter is more effective in resolving conflicts particularly in context aggregation and in general in performing the functions on the context distribution layer, as we have mentioned in Table 3.

In the final case conflict resolving policies applied at the context distribution layer to generate the events of interest for various subscribed applications, as it is nearly impossible for a human to analyze those context objects. Firstly, we only used the significance of context information to generate the events of interest as shown in Figure 5. We observed that only considering the significance of context objects is not sufficient to generate the events of interest. Therefore, we combined the trustworthiness of the source of context information with significance and found it quite useful to assist decision making in performing functions at the context distribution layer as shown in Table 3.

The overall behavior of our simulated experiments show that any QoC parameter alone is not sufficient to perform the task of decision making on any layer of context-

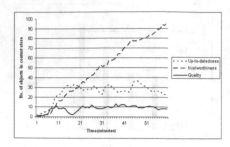
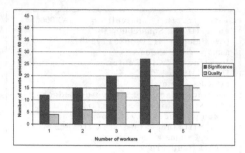

Fig. 4. Context aggregation using different QoC-based conflict resolving policies

Fig. 5. Events generated with help of QoC-based policies

Table 3. Table showing importance of different policies at different layers

Policy \ Layers	Context Acquisition	Processing	Context Distribution	Application
up-to-dateness based policy	+ +	+ +	+ +	+ + +
trustworthiness based policy	+ + +	+ +	+ +	+ + +
completeness based policy	+ +	+ +	+ +	+ + +
significance based policy	+	+ +	+ +	+ + +
combinations based policy	+ + +	+ + +	+ + +	+ + +

aware systems. QoC parameters used in combination of two or more parameters are more effective to perform this functionality at different layers of context-aware systems. We have also observed that the value of QoC parameters for different applications merely depends on the need of that specific application as shown in Table 3. In our experiments we gave preference to context objects with greater values of the QoC parameter over context objects with lower values of the QoC parameter. There can also be the situations where more sophisticated reasoning is required to make decisions on the basis of these parameters. These reasoning can be made on the basis of probability theories such as Basian theory, Dempster-Shafer theory or on the basis of neural networks.

6   Related Work

Different policies have been defined in literature to resolve the conflicts in context-aware systems. Mostly these policies are based on involving the user in the mediation process [7], resolving the conflicts by using some predefined static policies based on user preferences [17], discarding all the conflicting context, discarding the last received, or discarding the first received context objects [24]. Some works have also used QoC parameters to perform different tasks in middleware solutions to manage context information. In the remaining section we will discuss about the works that have suggested different conflict resolving policies and have used QoC parameters to perform different tasks in context-aware systems.

6.1 Conflict Resolving Policies

Xu et al. in [24] presented an impact-oriented automatic resolution of pervasive context inconsistency. Trying all possible resolution policies to find one that brings the least impact on the context-awareness of applications is referred as impact-oriented resolution. Conflict resolution scheme such as drop-latest, drop-all and drop-earliest are used in this work. But it had not presented the procedure for the situation evaluation of a context-aware system to calculate the impact of a resolution strategy.

A model for managing context information and resolving the inconsistencies is presented by Bu et al. in [2]. This work has used the policy of discarding the context with a low value of relative frequency in conflicting cases. Accepting new data and discarding the existing conflicting context object or rejecting the new data and keeping the old conflicting context object are used as accept and reject conflict resolving policies in [23]. But there has not been any assurance whether the new or old data is more reliable or not. Both of these policies can result in discarding the important context information. Dey et al. [7] involved the user in the mediation process to resolve ambiguity in context information.

In [17], Park et al. suggested to resolve conflicting situations between the applications by using a static policy based on user preferences that describes how conflicting applications need to adapt in case of conflicts between them. A configuration file gives a set of rules that specifies the behavior of applications in different contexts. These static policies may not comply to user needs in more dynamic and unknown environments, as Capra et al. argued in [4] context conflicts cannot be resolved at the design time and need to be resolved at the execution time.

6.2 Using QoC Parameters

Mihaila et al. [16] have identified four quality of data parameters: completeness, recency, frequency of updates, and granularity and have used these parameters to make source selection and ranking on WWW. Sources publish these quality of data parameters using WS-XML and they propose a query language that exploits those quality of data parameters to make source selections and rankings on the WWW. Chantzara et al. [5] have presented an approach that used quality of information for evaluating and selecting the information to be used as context information. They calculate a utility function based on QoC attributes.

Huebscher et al. [9] have also used QoC parameters in their adaptive middleware for context-aware applications in smart homes. They have used QoC parameters to perform different tasks in their middleware, such as context provider selection. In [15], we have used QoC parameters to detect and remove duplicate and conflicting information and perform context aggregation. Sheikh et. al. [19] have also used QoC parameters to enforce privacy of a user. But all these works [16, 5, 9, 19] are based on the assumption that sources of information also provide information about the quality parameters. In this case sources of information can affect the decisions based on those quality parameters. In contrast to this approach, we evaluate QoC parameters in our quality-aware context information management framework that works as a middleware solution to provide context information to context-aware applications and users.

7   Conclusion and Future Work

In this paper we have discussed conflicting situations that can occur at different layers of a context management system. We have also presented the conflict resolving policies based on QoC parameters that can be used to resolve the conflicts in such situations. We have performed the experiments to evaluate the performance of different policies and observed that conflict resolving policies that are defined upon the combination of different QoC parameters considering the context of the use of context information by a specific application showed better performance. For our next steps, we plan to use these policies to do more sophisticated reasoning in the fusion of low level context and extraction of high level context information. We also plan to enhance the quality of context information by combining the context information and QoC parameters from more than one context objects.

References

1. Baldauf, M., Dustdar, S., Rosenberg, F.: A survey on context-aware systems. Int. J. Ad Hoc Ubiquitous Comput. 2(4), 263–277 (2007)
2. Bu, Y., Gu, T., Tao, X., Li, J., Chen, S., Lu, J.: Managing quality of context in pervasive computing. In: QSIC 2006: Proceedings of the Sixth International Conference on Quality Software, pp. 193–200. IEEE Computer Society, Los Alamitos (2006)
3. Buchholz, T., Küpper, A., Schiffers, M.: Quality of context: What it is and why we need it. In: Proceedings of the 10th International Workshop of the HP OpenView University Association(HPOVUA). Hewlet-Packard OpenView University Association (2003)
4. Capra, L., Emmerich, W., Mascolo, C.: Carisma: Context-aware reflective middleware system for mobile applications. IEEE Transactions on Software Engineering 29, 929–945 (2003)
5. Chantzara, M., Anagnostou, M., Sykas, E.: Designing a quality-aware discovery mechanism for acquiring context information. In: AINA 2006: Proceedings of the 20th International Conference on Advanced Information Networking and Applications. AINA 2006, vol. 1, pp. 211–216. IEEE Computer Society, Los Alamitos (2006)
6. Cook, D.J.: Cook. Making sense of sensor data. IEEE Pervasive Computing 6(2), 105–108 (2007)
7. Dey, A.K., Mankoff, J.: Designing mediation for context-aware applications. ACM Trans. Comput.-Hum. Interact. 12(1), 53–80 (2005)
8. Huebscher, M.C., McCann, J.A., Dulay, N.: Fusing multiple sources of context data of the same context type. In: Proceedings of the 2006 International Conference on Hybrid Information Technology, pp. 406–415. IEEE Computer Society, Los Alamitos (2006)
9. Huebscher, M.C., McCann, J.A.: Adaptive middleware for context-aware applications in smart-homes. In: MPAC 2004: Proceedings of the 2nd workshop on Middleware for pervasive and ad-hoc computing, pp. 111–116. ACM, New York (2004)

10. Juszczyk, L., Psaier, H., Manzoor, A., Dustdar, S.: Adaptive query routing on distributed con-text - the cosine framework. In: International Workshop on the Role of Services, Ontologies, and Context in Mobile Environments (ROSOC-M), 10th International Conference on Mobile Data Management (MDM 2009), Taipeh, Taiwan, May 18-20. IEEE Computer Society Press, Los Alamitos (2009)
11. Kara, N., Dragoi, O.A.: Reasoning with contextual data in telehealth applications. In: Proceedings of the Third IEEE International Conference on Wireless and Mobile Computing, Networking and Communications. IEEE Computer Society, Los Alamitos (2007)
12. Kim, Y., Lee, K.: A quality measurement method of context information in ubiquitous environments. In: ICHIT 2006: Proceedings of the 2006 International Conference on Hybrid Information Technology, pp. 576–581. IEEE Computer Society, Los Alamitos (2006)
13. Krause, M., Hochstatter, I.: Challenges in modeling and using quality of context (qoc). In: Magedanz, T., Karmouch, A., Pierre, S., Venieris, I.S. (eds.) MATA 2005. LNCS, vol. 3744, pp. 324–333. Springer, Heidelberg (2005)
14. Manzoor, A., Truong, H.L., Dustdar, S.: On the evaluation of quality of context. In: Roggen, D., Lombriser, C., Tröster, G., Kortuem, G., Havinga, P. (eds.) EuroSSC 2008. LNCS, vol. 5279, pp. 140–153. Springer, Heidelberg (2008)
15. Manzoor, A., Truong, H.L., Dustdar, S.: Quality aware context information aggregation system for pervasive environments. In: The 5th International Symposium on Web and Mobile Information Services, The IEEE 23rd International Conference on Advanced Information Networking and Applications (AINA 2009), Bradford, UK, May 26-29. IEEE Computer Society, Los Alamitos (2009)
16. Mihaila, G.A., Raschid, L., Vidal, M.e.: Using quality of data metadata for source selection and ranking. In: Vossen (ed.) Proceedings of the Third International Workshop on the Web and Databases, WebDB 2000, AdamÕs Mark Hotel, pp. 93–98 (2000)
17. Park, I., Lee, D., Hyun, S.J.: A dynamic context-conflict management scheme for group-aware ubiquitous computing environments. In: COMPSAC 2005: Proceedings of the 29th Annual International Computer Software and Applications Conference (COMPSAC 2005), vol. 1, pp. 359–364. IEEE Computer Society, Los Alamitos (2005)
18. Perich, F., Joshi, A., Finin, T.W., Yesha, Y.: On data management in pervasive computing environments. IEEE Trans. Knowl. Data Eng. 16(5), 621–634 (2004)
19. Sheikh, K., Wegdam, M., van Suinderen, M.: Quality-of-context and its use for protecting privacy in context aware systems. Journal of Software 3, 83–93 (2008)
20. The EU WORKPAD Project, http://www.workpad-project.eu
21. Wu, H., Siegel, M., Ablay, S.: Sensor fusion using dempster-shafer theory ii: static weighting and kalman filter-like dynamic weighting. In: Proceedings of the 20th IEEE Instrumentation and Measurement Technology Conference, 2003, pp. 907–912 (2003)
22. Wun, A., Petrovi, M., Jacobsen, H.-A.: A system for semantic data fusion in sensor networks. In: DEBS 2007: Proceedings of the 2007 inaugural international conference on Distributed event-based systems, pp. 75–79. ACM, New York (2007)
23. Xu, C., Cheung, S.C.: Inconsistency detection and resolution for context-aware middleware support. In: Proceedings of the 10th European software engineering conference, pp. 336–345. ACM, New York (2005)
24. Xu, C., Cheung, S.C., Chan, W.K., Ye, C.: On impact-oriented automatic resolution of pervasive context inconsistency. In: Proceedings of the the 6th joint meeting of the European software engineering conference and the ACM SIGSOFT symposium on The foundations of software engineering, pp. 569–572. ACM, New York (2007)

Presentation and Evaluation of Inconsistencies in Multiply Represented 3D Building Models

Michael Peter

Institute for Photogrammetry, Universitaet Stuttgart,
Geschwister-Scholl-Str. 24D, 70174 Stuttgart
michael.peter@ifp.uni-stuttgart.de

Abstract. Open architectures demand for a federation of data from different context providers, which nearly always will be inconsistent to a certain degree. We present an approach for the evaluation and presentation of inconsistencies in multiply represented 3D building models and provide means for the minimization of ground plan inconsistencies. The presented approaches are tested using differently detailed models from various sources.

Keywords: evaluation, inconsistency, 3D, city model, adjustment.

1 Introduction

The increasing variety of applications which are based on spatial information resulted in a tremendously growing need for geospatial data. These demands are traditionally fulfilled by commercial vendors or governmental authorities, meanwhile also user-generated content like Open Street Map becomes more and more popular [1]. Since the data is captured by different providers, one object of the landscape was for example captured at different acquisition times, with different quality characteristics and different scales. Additionally it is stored in several databases and in different data models. This results in highly inconsistent data bases. The required integration of such multiple representations is a major research challenge in the field of GIS. Existing approaches mainly aim at the integration of 2D geospatial databases like street maps [2] or the evaluation of generalized 2D buildings [3]. However, we are aiming at the evaluation of inconsistencies in multiply represented 3D data as it is required for the processing of 3D building models used for applications in the context of urban planning, tourism, real estate presentation or personal navigation. Hence, these different purposes result in considerable differences with respect to the amount of detail or geometric accuracy for the available data.

Within the following section, our evaluation approach is presented based on different 3D building models covering the city of Stuttgart. Section 3 describes a first approach to minimize such detected inconsistencies. This is exemplarily implemented by an adjustment of 3D building models of relatively low geometric accuracy and small amount of detail to existing ground plans which were captured at a better geometric quality.

K. Rothermel et al. (Eds.): QuaCon 2009, LNCS 5786, pp. 156–163, 2009.

2   Inconsistency Evaluation

In this section, we describe our approach for the evaluation and presentation of inconsistencies between differently detailed 3D building models. This is done by comparing faces in both the reference and the input model that are equal in type. Relevant faces in the input model are projected into the coordinate system defined by the reference model's face and the intersection is computed. The ratio between the sum of all relevant faces' intersections and the reference face's area together with the mean angle and mean distance in between are mapped to the interval [0;1] and used to colour the input face.

2.1   Test Data

In order to test the consistency evaluation approach, we use differently detailed data from four different sources. Thus, our test data consists of very detailed 3D building models from terrestrial data collection, an area covering data set from airborne photogrammetric measurements, a generalised city model derived from this area covering data set and extruded building outlines from Open Street Map.

The data set providing the highest level of detail was collected by order of the City Surveying Office of Stuttgart using terrestrial measurements. This data set features hand-crafted models of landmarks and photo textured facades of the main part of Stuttgart downtown. It was collected for selected buildings of Stuttgart aiming not only for an internal use in city planning scenarios, but also for visualization purposes as for example in Google Earth.

The next available level of detail is a city model, which is available area covering from airborne photogrammetric data collection. This medium detailed model was constructed combining existing ground plans from cadastral maps as provided by the City Surveying Office and roof shapes reconstructed from aerial images [4].

The third data source consists of 3D building models, which were derived from the aforementioned medium detailed data set by the generalization approach described in Kada [5]. This algorithm aims to reconstruct a simplified representation of the input building model by means of searching the main planes of the original model and subtending these planes in order to build a correct boundary representation. However, the models evolving from this approach differ from the original building models due to the averaging operation as it is implemented in the simplification process.

The finally used Open Street Map (OSM) data is expected to be the least detailed and least accurate source of information, caused by its acquisition method. Aiming to be a free and open source alternative to commercial map services, the complete Open Street Map consists of user-generated content. For its acquisition, volunteers mostly use consumer GPS receivers or copy points of interest, streets or building outlines from aerial photos released by their owners. While this map currently only contains 2D data, access to additional sensors and straightforward modeling tools may allow for user-generated 3D building models in the future. For selected building ground plans from this data set, the WGS84-coordinates were transformed to the German Gauß-Krüger coordinate system and the ground plans were extruded to the eaves height of the model taken from the official Stuttgart city model, resulting in 3D block models similar to those constructed with the approach described in [6].

2.2 Evaluation Approach

Our approach to evaluate the differences between two building models is based on the analysis of the respective faces as they are available in the so-called reference and the input model.

For every face in the usually highly detailed and accurate reference model, a local coordinate system is constructed. In the case of horizontal faces, this is the face's normal vector and its cross product with the x-axis of the model coordinate system, complemented to a right-hand-system. For all other faces, the z-axis is used instead of the x-axis.

Input model faces relevant for the comparison to the currently evaluated reference model face are compiled according to their type, where a distinction between wall and roof faces is made. Then, this set of faces is further downsized by comparing the normal vectors. However, instead of using an angular threshold, only faces with opposite direction to the reference face are removed as these are not likely to represent a similar building feature.

The relevant faces are then projected into the local coordinate system and the intersection of the actual reference face and the projected relevant face is computed using the General Polygon Clipper library [7]. If an intersection polygon exists, its area is computed. However, faces exceeding a distance threshold with their mean distance to the reference face are excluded. This is necessary, as for the final consistency value distance and angular inconsistencies will be merged with the areal differences. Faces exceeding the distance threshold are nevertheless regarded in the consistency computation by their missing area.

The final consistency value per face is computed as

$$c_r = \frac{\sum_i \left(1 - \frac{d_i}{d_{max}} - |\alpha_i|\right) \cdot A_i}{A_r} \tag{1}$$

with d_i being the mean distance and α_i the mean angle between face i and the reference face, A_i being the area of the respective input face and A_r the area of the reference face.

This value in the interval $[0;1]$ may then be used in the visualization process. When used for example in $RGB = [1, c_r, c_r]$, the inconsistency of the input model to the current reference face is coloured from white (meaning maximum consistent) to red (maximum inconsistent). Results using the test data presented in the next section can be seen in the figures in section 2.3.

2.3 Results

As we consider the new Stuttgart city model the most accurate and detailed, we use the models stemming from this data source as the reference in our inconsistency evaluations.

In figure 1, the results for the Rosenstein museum models can be seen. As expected, the OSM model differs quite strongly from the reference model. However, the bigger differences in the longer walls in contrast to medium inconsistencies in the shorter sides reproduce quite well the shift of the complete building model, which can be seen when comparing the models with the naked eye. According to the OSM accuracy evaluation carried out by [8], differences in this range are to be expected. However, the areal differences used in the inconsistency evaluation may be too optimistic as the 3D wall faces evolve from the eaves heights of the Stuttgart city model and therefore are very similar to this model (see section 2.1).

In the generalized model, the strongly simplified roof structure shows the most distinct inconsistency to the reference model, with slight differences for the atrium and flat roof sides. The inconsistencies in the wall planes are mainly due to averaging during the generalization process and may therefore be minimized using the approach presented in chapter 3.

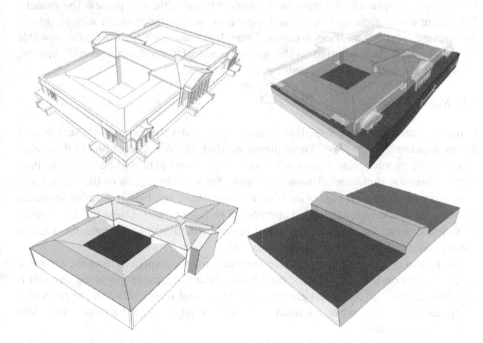

Fig. 1. Clockwise: Inconsistencies of OSM-model, generalized model and city model from airborne data collection in comparison to the city model from terrestrial data collection (upper left), coloured according to section 2.2

The city model from airborne data collection, however, holds high consistency in the main wall planes. As both of these models are provided by the city surveying office, this is most likely due to the shared data basis and accurately measured ground plans. The slight inconsistencies in the roof planes stem mainly from differently modelled roof angles, whereas the atrium without a match in the model from airborne data acquisition is marked clearly visible.

Figure 1 therefore illustrates the level of detail improving from OSM to the city model from airborne data collection. While most of the inconsistencies evolve from these differing levels of detail, the OSM as well as the generalized model show additional ground plan inconsistencies, which may be minimized by the algorithm presented in the next chapter.

3 Minimization of Ground Plan Inconsistencies

In the following, we describe our approach for the adjustment of less detailed building models, which are used as input models to the ground plans of higher detailed and more accurate models, which for our algorithm provide the reference models. Its main idea is the description of the input model subject to movable wall planes. The model's 3D structure is represented by the decomposition into distance ratios with respect to the movable planes and fixed z-values. Using least squares adjustment, the movable wall planes are then adjusted to the major planes of the reference model, causing changes to the faces depending on them.

3.1 Model Analysis

In the first step, the faces of the less detailed input model are merged to planes using a distance and angle threshold. These planes are then classified according to their adjacency to the ground plan. Planes adjacent to the ground plan will be shifted in their normal direction during adjustment. To ensure for a minimization of the inconsistencies between lesser and higher detailed representation, the higher detailed reference model is analyzed in order to find appropriate shifting targets. Using the faces' areas as weights, these are constructed as the planes with maximum weight for a set of parallel faces below a given distance threshold.

In order to adjust the input planes to the major planes computed from the reference model, correspondences have to be established. Therefore, the major plane's weight is weighed against the distance between input plane and major plane in the form of a computed ratio. The respective input plane will be adjusted to the major plane with maximum ratio value.

In order to adjust the complete model to the major planes of the reference model, the remaining building structure has to be decomposed into parameters suitable to describe it subject to the wall planes adjacent to the ground plan. In the case of sloped roof planes, this is done by computing the distance ratios in the xy-plane shown in figure 2.

To avoid topological errors evolving for example from changes in the ridge and eave lines, the slope of these roof planes will be changed during adjustment, which is established by maintaining the z-value of the ridge line as well as the z-difference between ridge and eave line. For wall planes not adjacent to the ground plan, similar distance ratios are used, while flat roof planes are left unchanged.

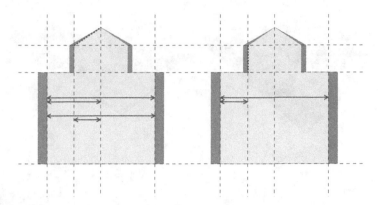

Fig. 2. Distance ratios for planes not adjacent to ground plan (left: roof planes, right: wall planes; dark grey: situation before adjustment)

3.2 Least Squares Adjustment

The final model is obtained using least squares adjustment. In order to maintain characteristics like rectangularity and parallelism, the planes adjacent to the ground plan are merely shifted minimizing the distances to the resampled intersection lines between major planes and ground plane. This simplified 2D approach is applicable under the assumption of vertical wall planes. The reconstruction of the remaining 3D structure based on the distance ratio values computed before and the fixed height values completes the adjustment process.

3.3 Results

In figure 3, the result of the adjustment can be seen. Here, the city model from airborne photogrammetric acquisition was chosen as the reference model and the OSM respectively the generalized model were adjusted to it. Using the inconsistency evaluation approach from section 2, remarkable differences can be seen, particularly when using the OSM model. Besides minimized ground plan inconsistencies, the adjustment may also help in restoring symmetric structures which were affected by the generalization, as visible at the New Castle model's side wings.

The presented adjustment approach only works reliably, if all roof faces in the input model can be described by two wall faces similar in the projected direction and the associated parameters. Otherwise, these roof faces will not be adjusted at all which may lead to topological errors in the resulting model. Thus, highly detailed models like the ones taken directly from the two city models, may not serve as input models. In contrast, as they originate from official sources, they are rather considered as reference models for the adjustment.

As this approach reduces the 3D adjustment problem to two dimensions, it may also be used if the model considered more accurate only consists of a 2D ground plan, allowing for the adjustment of arbitrary building models to accurately measured ground plans.

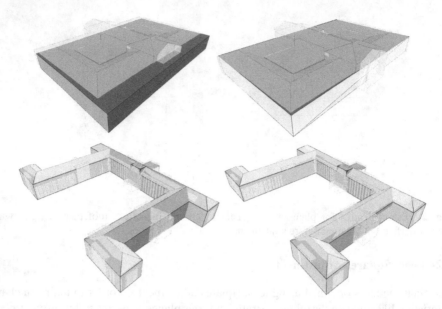

Fig. 3. First row: Rosenstein OSM model before (left) and after (right) adjustment to city model from airborne data collection (transparent); second row: New Castle model generalized (left) and adjusted (right); coloured according to section 2.2

4   Conclusions and Outlook

Within the paper, an approach for the automatic detection of geometric inconsisten-cies between multiple representations of 3D city models is presented. Also based on such detected inconsistencies an adjustment process is used to combine these 3D data sources of different quality. One scenario for the usage of both approaches would be community-based change detection. As communities like Open Street Map rely on heavily distributed observations, their data is very likely more up-to-date than that from governmental or commercial sources. Detected inconsistencies could be used to initiate local revisions, while the version from the community data is visualized, ad-justed to the ground plan in order to avoid errors related to building models in the vicinity.

Currently the analysis is based on a relatively simple distance measurement be-tween the respective building parts. The implementation of a more advanced evalua-tion, which could also include a topological analysis and semantic attributes will be part of our future work.

Acknowledgements

The research described in this paper is founded by "Deutsche Forschungsgemein-schaft (DFG, German Research Foundation). It takes place within the Collaborative Research Centre No. 627 "NEXUS – SPATIAL WORLD MODELS FOR MOBILE

CONTEXT-AWARE APPLICATIONS" at the Universitaet Stuttgart. The 3D building models are provided by Stadtmessungsamt Stuttgart, which is gratefully acknowledged. The Rosenstein Open Street Map ground plan was created by the users xylome2, BerndR, CvR and LayBack.

References

1. Haklay, M., Weber, P.: OpenStreetMap: User-Generated Street Maps. IEEE Pervasive Computing 7(4), 12–18 (2008)
2. Volz, S., Walter, V.: Linking different geospatial databases by explicit relations. In: Proceedings of the XXth ISPRS Congress, Comm. IV, Istanbul, Turkey, pp. 152–157 (2004)
3. Filippovska, Y., Walter, V., Fritsch, D.: Quality Evaluation of Generalization Algorithms. In: The International Archives of the Photogrammetry, Remote Sensing and Spatial Information Sciences. Beijing, China, p. 799
4. Wolf, M.: Photogrammetric data capture and calculation for 3D city models. In: Photogrammetric Week 1999, pp. 305–312 (1999)
5. Kada, M.: Scale-Dependent Simplification of 3D Building Models Based on Cell Decomposition and Primitive Instancing. In: Winter, S., Duckham, M., Kulik, L., Kuipers, B. (eds.) COSIT 2007. LNCS, vol. 4736, pp. 222–237. Springer, Heidelberg (2007)
6. Neubauer, N., Over, M., Schilling, A., Zipf, A.: Virtual Cities 2.0: Generating web-based 3D city models and landscapes based on free and user generated data (OpenStreetMap). In: GeoViz 2009. Contribution of Geovisualization to the concept of the Digital City. Workshop. Hamburg, Germany (2009)
7. Murta, A.: GPC General Polygon Clipper library [Internet] [zitiert 2009 Mai 25], http://www.cs.man.ac.uk/~toby/alan/software/
8. Haklay, M.: How Good is OpenStreetMap Information? A Comparative Study of OpenStreetMap and Ordnance Survey Datasets for London and the Rest of England. Environment & Planning (under review) (2008)

Bringing Quality of Context into Wearable Human Activity Recognition Systems

Claudia Villalonga[1,2], Daniel Roggen[1], Clemens Lombriser[1], Piero Zappi[3], and Gerhard Tröster[1]

[1] Wearable Computing Lab., ETH Zürich, Switzerland
{villalonga,droggen,lombriser,troster}@ife.ee.ethz.ch
[2] SAP Research, CEC Zürich, Switzerland
[3] DEIS, University of Bologna, Italy
pzappi@deis.unibo.it

Abstract. Quality of Context (QoC) in context-aware computing improves reasoning and decision making. Activity recognition in wearable computing enables context-aware assistance. Wearable systems must include QoC to participate in context processing frameworks common in large ambient intelligence environments. However, QoC is not specifically defined in that domain. QoC models allowing activity recognition system reconfiguration to achieve a desired context quality are also missing. Here we identify the recognized dimensions of QoC and the performance metrics in activity recognition systems. We discuss how the latter maps on the former and provide provide guidelines to include QoC in activity recognition systems. On the basis of gesture recognition in a car manufacturing case study, we illustrate the signification of QoC and we present modeling abstractions to reconfigure an activity recognition system to achieve a desired QoC.

1 Introduction

Context-awareness [1] allows applications or systems to adjust their behavior according to user state and needs. In wearable computing it leads to smart-assistants, e.g. in industrial applications [2] or healthcare [3]. In pervasive environments it enhances the experience of the user and allows advanced services, e.g. in elderly support [4], hospitals [5], or interactive museums [6]. It is also used to design more user friendly or personalized software applications [7,8].

Context is usually imperfectly known [9] due to e.g. sensor failures, erroneous sensor readings, wrong classification or user variability. Quality of Context (QoC) (or Quality of Context Information, QoI or QoCI), indicates the degree to wich the context reflects the real world [10]. Context reasoning and decision making may be improved by taking QoC into account, and context-aware actuation may be enacted only when sufficient certainty about the context is assessed. QoC-aware frameworks are investigated to enable large scale context-sensing and reasoning, in particular with respect to future Internet initiatives aiming at a networked context-aware physical world [11].

K. Rothermel et al. (Eds.): QuaCon 2009, LNCS 5786, pp. 164–173, 2009.

Context in wearable computing is inferred from on-body sensors by machine learning techniques. Here we consider activity and gesture recognition that is a particularly important aspect of context for industrial or healthcare applications. QoC in wearable activity recognition systems has not been studied in depth so far, only [12] goes in this direction by proposing a quality analysis algorithm running in parallel to the actual recognition. Metrics like accuracy reflect recognition performance, but QoC has not been explicitly applied to activity recognition. Diverging goals explain this, QoC modelling tends to be generic and domain independent; performance of activity recognition system is domain specific and derives much of the metrics from the machine learning field.

Due to the ever increasing number of sensors in ambient intelligence environments, there is also nowadays chance for context recognition systems to adjust the information processing path (e.g. by selecting best algorithms, sensors and sensor combinations) in order to reach a desired QoC. QoC thus represents a function that may be optimized by adjusting the context recognition chain with an appropriate model linking performance parameters to QoC.

The scope of this paper is twofolds. First, we make the link between QoC and performance metrics in wearable activity recognition systems. We summarize commonly recognized QoC parameters. On the basis of a review of wearable activity recognition scenarios, we identify the common performance metrics. We then show how performance metrics map onto QoC and discuss how the QoC parameters can be calculated in function of the performance metrics. In a broader sense, we propose guidelines to bring performance parameters of wearable activity recognition systems into a context framework that allows tagging context with complex QoC metadata. Second, we show how activity recognition systems' performance can be modeled. We describe three modeling approaches of various abstractions and illustrate how they apply to activity recognition systems. We outline how these models could be used to reconfigure an activity recognition system to achieve a desired QoC.

2   Quality of Context

Imperfect context information can influence an application's decisions and lead to wrong conclusions in some of the context processing elements and even the whole application. Quality of Context (QoC) counters these effects by adding parameters to the context which indicate to which extend the context data corresponds to the real world. Buchholz et al. define QoC as *"any information that describes the quality of information that is used as context information. Thus, QoC refers to information and not to the process nor the hardware component that possibly provide the information"* [10]. Considering that QoC only depends on the piece of context it relates to, in our opinion, the QoC value is associated by the context source and must not be modified during the information lifetime. This implies that all applications receive the same context information with the same QoC value and to check if the quality fits their necessities, they weight the QoC parameters according to their relevance or importance.

We present several QoC parameters that have been identified in literature and give a unified name to parameters describing the same concepts:

- **Measurement time** – Time of the last context attribute value measurement [9,10,13,14,15,16,17,18].
- **Delay time** – Interval between the time a situation occurs in the real world and the time the measured context attribute value becomes available to the system [19].
- **Temporal Scope** – Period of time a context attribute value is valid after its measurement [14,17].
- **Origin** – Identifier of the sensor, processor or context source that provided the measurement of the context attribute [16,18].
- **Spatial Scope** – Physical or virtual area or domain within which the context attribute value is valid [17].
- **Resolution** – Granularity or minimal perceivable change of the context attribute value measurement [10,15,17,20].
- **Accuracy** – Degree of closeness of the context attribute value to the real world situation [9,14,18,20,21].
- **Probability of correctness** – Probability of the context attribute value matching the real world situation [10,14,15,17,19,20].

QoC ougth to be linked to contextual information delivered to an application consuming context. It must thus be included in context models and handled by context frameworks. Examples of ontologies including QoC are [15,19,22], while [9] provides QoC parameters in their graphical model.

A framework fully supporting QoC must include it in the collection, interpretation, access and delivery of context information, as well as other functionalities to efficiently handle QoC, e.g. resolving conflicts using QoC or only delivering context with a minimum QoC. Some examples of such frameworks for small scale and centralized scenarios are presented in [9,13,14].

3   Performance Metrics for Activity Recognition

Recognizing human activities and manipulative gestures from body-worn sensors is essentially a classification problem, where machine learning techniques map sensor patterns to activities or gestures. We surveyed activity recognition systems in various scenarios including: industrial assistance [2], sports [23], HCI [24], health [25], and others [26,27,28,29,30]. We identified the common parameters describing human activity recognition systems performance. Essentially, they relate to the metrics used to characterize the underlying machine learning techniques.

Accuracy – The number of correctly recognized events divided by the total number of events. It is commonly used as a single number to describe the performance. However, it provides little insight into what kind of errors occur.

Confusion matrix – Matrix of recognized activities versus the true activities (see Fig. 1(a)). The matrix diagonal contains the correctly recognized activities

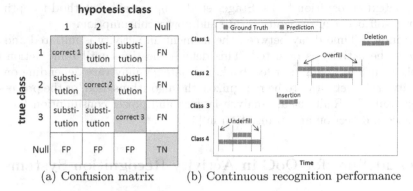

(a) Confusion matrix   (b) Continuous recognition performance

Fig. 1. Performance metrics of activity recognition a) confusion matrix and b) continuous recognition performance

(true positives) while the off-diagonal elements denote activities that have been confused with another activity. The off-diagonal elements can be summed up along the recognized activities row to obtain the false positives and along the true class column to obtain the false negatives.

Precision, Recall, and Specificity – The confusion matrix renders metrics like precision (true positives divided by positives), recall or sensitivity (true positives divided by true positives and false negatives), and specificity (true negatives divided by true negatives and false positives). These metrics are better suited than accuracy for continuous recognition systems [30], where only a subset of all possible user activities can be recognized and everything during the rest of the time is classified as "nothing happened", or the *NULL* class.

Continuous recognition performance – Ward [30] analyzes the errors occurring in continuous (or online) recognition and distinguishes between *insertions* for events that have been recognized, but have not happened in reality and *deletions* for events that happened in reality, but have not been recognized. *Merges* occur when multiple events are recognized as only one and *fragmentations* when single real events are split into multiple recognized ones. Finally, *overfill* and *underfill* describe how well the duration of an event has been recognized. Fig. 1(b) shows the errors in continuous recognition.

ROC curve – The Receiver Operating Characteristic (ROC) is a graphical plot of the true positive rate versus false positive rate for a recognition algorithm when its discrimination threshold is varied. The ROC curve shows how a recognition algorithms trades off a high recall to find all interesting events, but introducing false positives, for a high precision, which suffers from a larger number of false negatives. A commonly used metric derived from the ROC curve is the *Area under Curve* (AUC) which characterizes an algorithm by indicating whether it is valuing more importantly true positives or true negatives.

Power consumption – Bharatula et al. [27] define a cost metric for activity recognition algorithms that measures the power consumed for classification to solve context recognition tasks. Stäger et al. [26] present a method to optimize the trade-off between recognition rate and power consumption.

Latency – Time delay between the moment an activity is initiated and the moment the activity is detected. This defines the time until some action can be taken, such as giving user feedback. The latency may vary depending on the diversity of the activities to be recognized, the algorithms used, or the processing power available. Reilly et al. analyze latency and power consumption for real-time speech detection and translation in [31].

4 Guidelines for QoC in Activity Recognition Systems

Activity recogntion in wearable computing tackles on-body systems of limited size which differ considerably from the higher level view of context aware applications and large scale context frameworks. However, user activity is a valuable piece of context and is worth to be made available to any application through context frameworks. By connecting to context frameworks the activity recognition systems could obtain additional data from environmental sensors and even incorporate them into the recognition chain.

Integrating the two systems leads to the question of how QoC is calculated in function of the performance metrics, i.e. how these metrics often derived from the machine learning field are mapped into the abstract QoC parameters and and how QoC should be extended to be useful in this area.

Offline Performance Metrics – Accuracy as part of the QoC is one of the most relevant parameters as it gives an idea of the relation between the context value and reality. In wearable computing, the corresponding metrics are the offline performance metrics, i.e. accuracy, confusion matrix, precision, recall and specificity. Even if "accuracy" is used in both domains, the concept is different since in activity recognition it is a statistical value saying how often the recognized class matches real class. As a first recommendation, we suggest using the accuracy of specific recognized classes, i.e. the value on the diagonal of the confusion matrix, as the quantification for the accuracy parameter in the QoC of the recognized class. Consequently the recognition system will not have a single overall value for the accuracy of the detected classes, but an individual value for each recognized class. If we wanted to provide a single global value for the recognition chain accuracy, then the global accuracy in the activity recognition performance metrics should be used, but this is not recommendable since we loose a lot of information that is contained in the confusion matrix. In fact, if we did not want to lose any information, then QoC should be re-defining to include the whole confusion matrix.

Online Performance Metrics – Activity recognition systems usually operate online. The continuous recognition performance metrics defined by [30] are particularly important to quantify the errors of the classification for this system. They however do not match on any recognized QoC parameters. We thus suggest extending QoC to include these performance metrics (insertions, substitutions, deletion, merge, fragmentation, overfill and underfill).

Cost of Context and Power Consumption – Context frameworks do not consider resource consumption when delivering or collecting context as the only goal is to deliver high quality context. In wearable computing, however, devices are running off batteries and only limited power is available. Thus it is important to consider how much power is invested into the activity recognition. Power consumption can be traded-off for accuracy [26] and can be used as performance metric of an activity recognition system. However, it is not a QoC measure as it does not indicate how the context represents the real world, but only informs about the cost to calculate this context value. Cost of Context is therefore a new concept, which we define as *"Cost of Context (CoC) is a parameter associated to the context that indicates the resource consumption used to measure or calculate the piece of context information"*.

Delay Time and Latency – In a large scale framework, the time to find the appropriate context source, processing the context information using e.g. ontology reasoning, and transmitting the resulting context over the network can range in the scope of seconds. In wearable computing scenarios involving human assistance, the response time of an application is crucial for the acceptance by the user. In some cases, feedback must be delivered within milliseconds for meaningful interaction. Therefore latency is another vital metric in human activity recognition and needs to be integrated into QoC. There is a clear matching between the latency and the delay time parameter of the QoC and we recommend to use the calculated latency as quantification for the delay time parameter of the QoC measure.

5 Modeling of QoC in Activity Recognition Systems

The value of QoC in complex activity recognition systems is influenced by variable parameters in the activity recognition chain, e.g. the accuracy and reliability of sensors, the sampling frequency of feature extraction block, the parameters of classifiers, the type of classifier or the methods used for fusion.

Allowing advanced functionalities, like the automatic creation and configuration of recognition chains which achieve a desired QoC, requires the modeling of the behavior of QoC for the recognized context in function of the characteristics of the activity recognition chain.

We introduce three levels of modeling for QoC and illustrate them on basis of the recognition chain depicted in Fig. 2. We used it in previous work [32] to detect the workers's activities in a car assembly manufacture. Activities are detected individually by redundant sensors using Hidden Markov Models (HMM) and fused over a group of nodes using a majority vote. We base the analysis for the accuracy parameter but the approach may equally apply to other QoC parameters.

Empirical Offline Modeling – A first step is to collect sensor data of the activities to be recognized and to annotate the data with a ground truth of when and for how long the activities were performed. This dataset can then be used to train and test the complete recognition chain offline. The relationship between the input parameters and the QoC of the output is therefore found empirically, i.e. by simulation. This approach can be applied if the activity recognition chain and the environment are static and well defined at design time.

The advantage of such a model is that a detailed understanding of the behavior of the system can be achieved by sweeping the parameters of interest at development time - the optimal parameters being used for implementation. The disadvantages of this approach is that often the resulting QoC is represented as an average for a given input parameter to simplify visualization.

Analytical Modeling – An analytical model involves deriving a mathematical expression for each element of the recognition chain such that it can provide the QoC of its output given the QoC of its inputs. As an example of the modeling we use the empirically determined characteristics of the classifiers and model it using a beta distribution. We then can use a model of a majority vote classifier to determine the output distribution of the system. The expected value $E[x']$ of the accuracy of majority voting with odd n voters given the expected value of the accuracy $E[x]$ of the classifiers generating the inputs behaves as follows [33]:

$$E[x'] = \sum_{k=\frac{k+1}{2}}^{n} \binom{n}{k} E[x]^k (1 - E[x])^{n-k}$$

An analytical model is fast and can be used for algorithm composition, such as to achieve guaranteed bounds. However, at runtime it provides only a statistical value which may lie considerably off the actual performance.

Empirical Online Modeling – In many cases, there is a large number of unknown or untested parameter combinations affecting real-world performance. E.g. sensors processing or sensor inclusion in the recognition may be changed dynamically. In such cases offline or analytical modeling according to all parameters (potentially not known at design time) may require too much memory or be too complex.

In this new approach we introduced in [32], QoC is assessed empirically, online, by keeping in memory a set of annotated reference instances for each sensors. When the recognition chain parameter changes these instances are passed through the chain and the resulting match with the ground truth indicates the

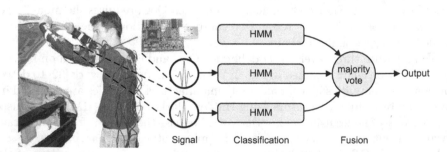

Fig. 2. Activity recognition to detect the workers's activities in a car assembly manufacture scenario. Data collected from the multiple redundant body-worn sensors is classified using Hidden Markov Models (HMM) and the individual classifications are fused using majority voting to obtain an improved decision.

QoC. We used this approach to select a group of sensors to achieve a desired QoC while optimizing network lifetime. Each time a sensor exhausted its battery it was replaced by one or more sensors to maintain the QoC [32].

The main advantages of this approach are that the performance of the system is found exactly, i.e. it is not an average statistical value and that it allows the dynamic addition of new features, classifier or classifier combinations. The disadvantages may be the slower QoC estimation and the larger memory needs.

6   Conclusion

We discussed the recognized QoC parameters and described the performance metrics in wearable activity recognition systems. We analyzed the links between these two domains and discussed the advantages of including activity recognition systems into larger scale QoC-aware context frameworks.

We discussed how the QoC parameters could be calculated for an activity recognition system and we identified that there is not a one to one mapping between QoC parameters and activity recognition performance metrics. Therefore, we propose the following to address these issues:

1. The values on the diagonal of the confusion matrix are used as the quantification for the accuracy parameter in the QoC of the recognized class.
2. The QoC is extended to include placeholders for metrics used in continuous activity recognition systems (insertions, substitutions, deletion, merge, fragmentation, overfill and underfill).
3. The Cost of Context (CoC) is added in parallel to the QoC measure to reflect the resource consumption necessary to calculate a piece of context.
4. The calculated latency is used as quantification for the delay time parameter of the QoC measure.

Finally we presented different approaches to model the QoC of a wearable human activity recognition system and showed how to link the recognition chain parameters to QoC. This modeling allows to define a tradeoff between Cost of Context and Quality of Context. In the future we will extend our modeling efforts to configure dynamically a recognition chain to achieve a desired Quality of Context taking into account the Cost of Context.

Acknowledgment

This paper describes work undertaken in the context of the projects SENSEI and OPPORTUNITY.

The SENSEI project, "Integrating the Physical with the Digital World of the Network of the Future" (www.sensei-project.eu), is a Large Scale Collaborative Project supported by the European 7th Framework Programme, contract number: 215923.

The project OPPORTUNITY acknowledges the financial support of the Future and Emerging Technologies (FET) programme within the Seventh Framework Programme for Research of the European Commission, under FET-Open grant number: 225938.

References

1. Dey, A.K., Abowd, G.D., Salber, D.: A conceptual framework and a toolkit for supporting the rapid prototyping of context-aware applications. Human-Computer Interaction 16(2), 97–166 (2001)
2. Stiefmeier, T., Roggen, D., Ogris, G., Lukowicz, P., Tröster, G.: Wearable activity tracking in car manufacturing. IEEE Pervasive Computing 7(2), 42–50 (2008)
3. Tentori, M., Favela, J.: Activity-aware computing for healthcare. IEEE Pervasive Computing 7(2), 51–57 (2008)
4. Consolvo, S., Roessler, P., Shelton, B., LaMarca, A., Schilit, B., Bly, S.: Technology for care networks of elders. IEEE Pervasive Computing 3(2), 22–29 (2004)
5. Bardram, J.E.: Applications of context-aware computing in hospital work: examples and design principles. In: Proc. ACM Symposium on Applied Computing (SAC), pp. 1574–1579 (2004)
6. Fleck, M., Frid, M., Kindberg, T., O'Brien-Strain, E., Rajani, R., Spasojevic, M.: From informing to remembering: Ubiquitous systems in interactive museums. IEEE Pervasive Computing 1(2), 13–21 (2002)
7. Schilit, B.N., Adams, N., Want, R.: Context-aware computing applications. In: Proc. IEEE Workshop on Mobile Computing Systems and Applications, pp. 85–90 (1994)
8. Böhm, S., Koolwaaij, J., Luther, M., Souville, B., Wagner, M., Wibbels, M.: Introducing IYOUIT. In: Proc. Int'l Semantic Web Conference, pp. 804–817 (2008)
9. Henricksen, K., Indulska, J.: Modelling and using imperfect context information. In: Proc. 2nd IEEE Conf. Pervasive Computing and Communications Workshops, pp. 33–37 (2004)
10. Buchholz, T., Kuepper, A., Schiffers, M.: Quality of context: What it is and why we need it. In: Proc. Workshop of the HP OpenView University Association, HPOVUA (2003)
11. SENSEI: http://www.sensei-project.eu/
12. Berchtold, M., Decker, C., Riedel, T., Zimmer, T., Beigl, M.: Using a context quality measure for improving smart appliances. In: Proc. 27th Int'l Conf. Distributed Computing Systems Workshops (ICDCSW), p. 52 (2007)
13. Lei, H., Sow, D.M., John, S., Davis, I., Banavar, G., Ebling, M.R.: The design and applications of a context service. ACM SIGMOBILE Mobile Computing Communications Review 6(4), 45–55 (2002)
14. Judd, G., Steenkiste, P.: Providing contextual information to pervasive computing applications. In: Proc. 1st IEEE Int'l Conf. on Pervasive Computing and Communications (PERCOM), p. 133 (2003)
15. Gu, T., Wang, X., Pung, H., Zhang, D.: An ontology-based context model in intelligent environments. In: Proceedings of Communication Networks and Distributed Systems Modeling and Simulation Conference, CNDS 2004 (2004)
16. Zimmer, T.: Qoc: Quality of context - improving the performance of context-aware applications. In: Advances in Pervasive Computing. Adj. Proc. Pervasive., vol. 207, pp. 209–214 (2006)
17. Sheikh, K., Wegdam, M., van Sinderen, M.: Middleware support for quality of context in pervasive context-aware systems. In: Proc. 5th IEEE Int'l Conf. on Pervasive Computing and Communications Workshops (PERCOMW), pp. 461–466 (2007)
18. Manzoor, A., Truong, H.L., Dustdar, S.: On the evaluation of quality of context. In: Roggen, D., Lombriser, C., Tröster, G., Kortuem, G., Havinga, P. (eds.) EuroSSC 2008. LNCS, vol. 5279, pp. 140–153. Springer, Heidelberg (2008)

19. Bu, Y., Gu, T., Tao, X., Li, J., Chen, S., Lu, J.: Managing quality of context in pervasive computing. In: Proc. 6th Int'l Conf. on Quality Software (QSIC), pp. 193–200 (2006)
20. Krause, M., Hochstatter, I.: Challenges in modelling and using quality of context (qoc). In: Magedanz, T., Karmouch, A., Pierre, S., Venieris, I.S. (eds.) MATA 2005. LNCS, vol. 3744, pp. 324–333. Springer, Heidelberg (2005)
21. Kim, Y., Lee, K.: A quality measurement method of context information in ubiquitous environments. In: Proc. Int'l Conf. on Hybrid Information Technology (ICHIT), vol. 2, pp. 576–581 (2006)
22. Strang, T., Linnhoff-Popien, C., Frank, K.: Cool: A context ontology language to enable contextual interoperability. In: Proc. 4th IFIP WG6.1 Int'l Conf. on Distributed Applications and Interoperable Systems (DAIS), pp. 236–247 (2003)
23. Heinz, E., Kunze, K., Gruber, M., Bannach, D., Lukowicz, P.: Using wearable sensors for real-time recognition tasks in games of martial arts – an initial experiment. Proc. IEEE Symposium on Computational Intelligence and Games, CIG (2006)
24. Kallio, S., Kela, J., Korpipää, P., Mäntyjärvi, J.: User independent gesture interaction for small handheld devices. Int'l J. of Pattern Recognition and Artificial Intelligence 20(4), 505–524 (2006)
25. Bächlin, M., Roggen, D., Plotnik, M., Hausdorff, J.M., Tröster, G.: Online detection of freezing of gait in parkinson's disease patients: A performance characterization. In: Accepted for Proc. 4th Int'l Conf. on Body Area Networks (2009)
26. Stäger, M., Lukowicz, P., Tröster, G.: Power and accuracy trade-offs in sound-based context recognition systems. Pervasive and Mobile Computing 3, 300–327 (2007)
27. Bharatula, N., Lukowicz, P., Tröster, G.: Functionality-power-packaging considerations in context aware wearable systems. Personal and Ubiquitous Computing 12(2), 123–141 (2008)
28. Van Laerhoven, K., Gellersen, H.W.: Spine versus porcupine: a study in distributed wearable activity recognition. In: Proc. 8th Int'l Symposium on Wearable Computers (ISWC), pp. 142–149 (2004)
29. Bao, L., Intille, S.S.: Activity recognition from user-annotated acceleration data. In: Ferscha, A., Mattern, F. (eds.) PERVASIVE 2004. LNCS, vol. 3001, pp. 1–17. Springer, Heidelberg (2004)
30. Ward, J., Lukowicz, P., Tröster, G., Starner, T.: Activity recognition of assembly tasks using body-worn microphones and accelerometers. IEEE Trans. Pattern Analysis and Machine Intelligence 28(10), 1553–1567 (2006)
31. Reilly, D., Siewiorek, D., Smailagic, A.: Power consumption and performance analysis of real-time speech translator smart module. In: Proc. 4th Int'l Symposium on Wearable Computers (ISWC), pp. 25–32 (2000)
32. Zappi, P., Lombriser, C., Farella, E., Roggen, D., Benini, L., Tröster, G.: Activity recognition from on-body sensors: accuracy-power trade-off by dynamic sensor selection. In: Verdone, R. (ed.) EWSN 2008. LNCS, vol. 4913, pp. 17–33. Springer, Heidelberg (2008)
33. Polikar, R.: Ensemble based systems in decision making. IEEE Circuits and Systems Magazine 6(3), 21–45 (2006)

Quality Dependent Reconstruction of Building Façades

Susanne Becker and Norbert Haala

Institute for Photogrammetry, Universitaet Stuttgart, 70174 Stuttgart, Germany
forename.lastname@ifp.uni-stuttgart.de

Abstract. The paper describes an approach for the quality dependent reconstruction of building façades using 3D point clouds from mobile terrestrial laser scanning. Due to different look angles, such measurements frequently suffer from different point densities at the respective building façades. In order to support the interpretation at areas, where no or only limited LiDAR measurements are available, a quality dependent processing is implemented. First, façades are reconstructed at areas of sufficient LiDAR point availability. Based on this reconstruction, rules are derived automatically, which together with the respective façade elements constitute a so-called façade grammar. It holds all the information which is necessary to reconstruct façades in the style of the given building. In our quality dependent approach, this grammar is used as knowledge base for the verification of a façade model reconstructed at areas of limited sensor data quality. Additionally, it is applied for the generation of synthetic façades where no LiDAR measurement is available.

Keywords: Architecture, Modelling, Interpretation, Building, Segmentation, 3D point clouds, laser scanning.

1 Introduction

Modelling and visualisation of 3D urban landscapes has been a topic of major interest in the past years. A number of tools for the production of virtual city models were developed, which are usually based on 3D measurements from airborne stereo imagery or LiDAR. In addition to this area covering airborne data collection, which mainly provides the outline and roof shape of buildings, terrestrial laser scanning (TLS) is frequently used, especially if a more accurate and detailed three-dimensional mapping of man-made structures is required. Beside measurements from fixed viewpoints, these scanners are often mounted on moving platforms. Such mobile mapping systems are usually also equipped with multiple video or digital cameras and allow for the rapid and cost effective capturing of 3D data for larger areas. This variety of sensors and platforms results in heterogeneous urban data sets of different accuracy, coverage and amount of detail. These differences in quality have to be considered during further processing.

Existing solutions are available for a purely geometric evaluation. One example is the automatic registration and transformation of the captured data sets to a reference coordinate system by so-called integrated georeferencing. For this purpose, least squares approximations are often used, which can also provide explicit information on

K. Rothermel et al. (Eds.): QuaCon 2009, LNCS 5786, pp. 174–184, 2009.

the underlying geometric data quality. However, such standard tools are not available if semantic information has to be derived for tasks like object reconstruction. One application in this context is the automatic extraction of façade structures for the generation of highly detailed 3D city models, which is the focus of our work. Generally, the reliability and accuracy of façade models derived from measured data depends on data quality in terms of coverage, resolution and accuracy. Façade parts for which only little or inaccurate 3D information is available cannot be reconstructed at all or require considerable manual pre- or postprocessing. In order to avoid such time-consuming user interaction, automatic algorithms for façade modelling which can cope with data of heterogeneous quality become important.

In our application, a formal grammar is applied for the explicit reconstruction of building façades using point clouds from terrestrial laser scanning. The main problem in this respect is the strong variation of the available point densities at the building façades, which frequently results from different look angles during the scanning process. We combine bottom-up and top-down modelling to handle this inhomogeneous data quality during reconstruction. The bottom-up modelling ensures flexibility to capture the great variety of real-world façade structures whereas the top-down modelling achieves topological correctness and robustness against potentially incomplete data sets of heterogeneous quality. While existing algorithms based on formal grammars still require manual interaction either for rule definition or façade interpretation, our algorithm runs fully automatically during all processing steps. The core of our façade reconstruction approach, thus, is the automatic generation of a formal grammar, which will be part of quality sensitive modelling strategies.

Our algorithm is implemented as follows: Firstly, rules are extracted automatically from observed façade geometries, which are - due to limitations during data acquisition - mostly available only for parts of a building. As discussed in section 2, these rules are represented by a so-called façade grammar. The rules then can be applied to generate façade structure for the remaining parts of the building. As demonstrated in section 3, we take advantage of this in various ways. Top-down predictions are activated and used for the verification and robustification of the reconstruction result that has already been derived from the observed measurements during the bottom-up modelling. Moreover, the façade grammar can be applied to synthesise façades for which no sensor data is available. Concerning the uncertainty of the generated façade models, first considerations on strategies for quality evaluation are presented in section 4. Exemplary 3D reconstruction results are given in section 5.

2 Grammar Based Modelling of Building Façades

Our algorithm starts with the bottom-up modelling of façade geometries using terrestrial LiDAR and image data as discussed in section 2.1. After this interpretation step, the resulting reconstructed façade serves as a knowledge base for further processing. Dominant or repetitive features and regularities as well as their hierarchical relationship are detected from the modelled façade elements. At the same time, production rules are automatically inferred. The rules together with the 3D representations of the modelled façade elements constitute a formal grammar which we will call *façade grammar*. It contains all the information which is necessary to reconstruct façades in the style of the respective building during top-down modelling. Section 2.2 will give a short overview of the use of formal grammars for the representation of building architecture. The automatic inference will be discussed in section 2.3.

2.1 Data Driven Reconstruction

The approach for the data driven façade reconstruction aims at refining an existing coarse building model by adding 3D geometries to the planar façades [1]. Windows, doors and protrusions are modelled from the LiDAR data by searching for holes in the point cloud measured at the building façade. These structures are then refined by integrating further 3D information derived from images of high resolution. The modelling process applies a 3D object representation by cell decomposition. The idea is to segment an existing coarse 3D building object with a flat front face into 3D cells. Each 3D cell represents either a homogeneous part of the façade or a window area. Therefore, they have to be differentiated based on the availability of measured LiDAR points. After this classification step, window cells are eliminated while the remaining façade cells are glued together to generate the refined 3D building model. The difficulty is finding planar delimiters from the LiDAR points that generate a good working set of cells. Since our focus is on the reconstruction of windows, the delimiters have to be derived from 3D points at the window borders by searching for holes in the point cloud. For the exemplary dataset "Alte Kanzlei, Stuttgart", Fig. 1a depicts the coarse building model with the aligned LiDAR points. Fig. 1b shows the refined fa-çade after reconstruction.

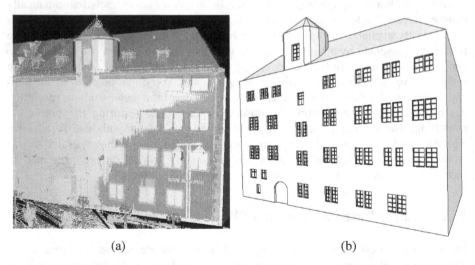

(a)   (b)

Fig. 1. Alte Kanzlei, Stuttgart: LiDAR points aligned with coarse 3D building model (a) and refined façade model (b)

2.2 Formal Grammars for the Modelling of Architecture

Usually formal grammars are applied during object reconstruction to ensure the plausibility and the topological correctness of the reconstructed elements. A famous example for formal grammars is given by Lindenmayer-systems (L-systems). Originally used to model the growth processes of plants, L-systems serve as a basis for the development of further grammars appropriate for the modelling of architecture. For instance, [2] produce detailed building shells without any sensor data by means of a shape grammar.

In our application, a formal grammar will be used for the generation of façade structure where only partially or no sensor data is available. In principle, formal grammars provide a vocabulary and a set of production or replacement rules. The vocabulary comprises symbols of various types. The symbols are called non-terminals if they can be replaced by other symbols, and terminals otherwise. The non-terminal symbol which defines the starting point for all replacements is the axiom. The grammar's properties mainly depend on the definition of its production rules. They can be, for example, deterministic or stochastic, parametric and context-sensitive. A common notation for productions which we will refer to in the following sections is given by

$$id: \quad lc < pred > rc : cond \quad \rightarrow \quad succ : prob$$

The production identified by the label *id* specifies the substitution of the predecessor *pred* for the successor *succ*. Since the predecessor considers its left and right context, *lc* and *rc*, the rule is context-sensitive. If the condition *cond* evaluates to true, the replacement is carried out with the probability *prob*. Based on these definitions and notations we develop a façade grammar which allows us to synthesise new façades of various extents and shapes. The axiom refers to the new façade to be modelled and, thus, holds information on the façade polygon. The sets of terminals and non-terminals, as well as the production rules are automatically inferred from the reconstructed façade as obtained by the data driven reconstruction process.

Existing systems for grammar based reconstruction of building models which derive procedural rules from given images or models still resort to semi-automatic methods [3], [4] and [5]. In contrast, we propose an approach for the automatic inference of a façade grammar in the architectural style of the observed building façade.

2.3 Automatic Inference of Façade Grammar

Based on the data driven reconstruction result, the façade grammar is automatically derived by searching for terminals, their interrelationship, and production rules.

Searching for terminals. In order to yield a meaningful set of terminals for the façade grammar, the building façade is broken down into some set of elementary parts, which are regarded as indivisible and therefore serve as terminals. For this purpose, a spatial partitioning process is applied which segments the façade into floors and each floor into tiles. Tiles are created by splitting the floors along the vertical delimiters of *geometries*. A geometry describes a basic object on the façade that has been generated during the data driven reconstruction process (section 2.1). It represents either an indentation like a window or a protrusion like a balcony or an oriel.

Two main types of tiles can be distinguished: wall tiles, which represent blank wall elements, and geometry tiles, which include structures like windows and doors. All these tiles are used as terminals within our façade grammar. In the remaining sections of the paper, wall tiles will be denoted by the symbols W for non-terminals and w_i for terminals. Geometry tiles will be referred to as G and g_i in case of non-terminals and terminals, respectively.

Interrelationship between terminals. Having distinguished elementary parts of the façade we now aim at giving further structure to the perceived basic tiles by grouping them into higher-order structures. This is done fully automatically by identifying hierarchical structures in sequences of discrete symbols. The structural inference reveals hierarchical interrelationships between the symbols in terms of rewrite rules. These rules identify phrases that occur more than once in the string. Thus, redundancy due to repetition can be detected and eliminated. As an example, Fig. 2a shows a modelled floor of the data set "Prinzenbau, Stuttgart". While Fig. 2b depicts the corresponding tile string in its original version, the compressed string and the extracted structures S_i (i=1,2,3) are given in Fig. 2c. The hierarchical relations between the façade elements can be stored in a parse tree illustrated in Fig. 2d.

a)
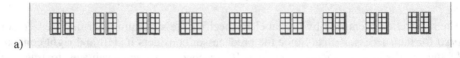

b) floor 1 $\rightarrow$ w_1 g_1 w_3 g_1 w_1 g_1 w_3 g_1 w_1 g_1 w_3 g_1 w_1 g_1 w_3 g_1 w_2 g_1 w_3 g_1 w_2 g_1 w_3 g_1 w_1 g_1
w_3 g_1 w_1 g_1 w_3 g_1 w_1 g_1 w_3 g_1 w_1

c)

floor 1 $\rightarrow$ w_1 S_3 w_2 S_1 w_2 S_3 w_1

$S_1 \rightarrow g_1\ w_3\ g_1$
$S_2 \rightarrow S_1\ w_1\ S_1$
$S_3 \rightarrow S_2\ w_1\ S_2$

d)

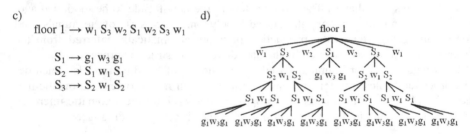

Fig. 2. Modelled floor of the data set "Prinzenbau, Stuttgart" (a), corresponding tile string (b), compressed tile string and extracted structures (c), parse tree (d)

Inference of production rules. Based on the sets of terminals $T=\{w_1, w_2, \ldots, g_1, g_2, \ldots\}$ and non-terminals $N=\{W, G, \ldots, S_1, S_2, \ldots\}$, which have been set up previously, the production rules for our façade grammar can be inferred. Following types of production rules are obtained during the inference process:

p_1: $F \rightarrow W+$
p_2: $W : cond \rightarrow W\ G\ W$
p_3: $G : cond \rightarrow S_i : P(x|p_3)$
p_4: $G : cond \rightarrow g_i : P(x|p_4)$
p_5: $lc < W > rc : cond \rightarrow w_i : P(x|p_5)$

The production rules p_1 and p_2 stem from the spatial partitioning of the façade. p_1 corresponds to the horizontal segmentation of the façade into a set of floors. The vertical partitioning into tiles is reflected in rule p_2. A wall tile, which in the first instance can stand for a whole floor, is replaced by the sequence wall tile, geometry tile, wall tile. Each detected structure gives rise to a particular production rule in the form of p_3. This rule type states the substitution of a geometry tile for a structure S_i. In addition, all terminal symbols generate production rules denoted by p_4 and p_5 in the case of geometry terminals g_i and wall terminals w_i, respectively.

3 Application – Knowledge Propagation

Our façade grammar implies information on the architectural configuration of the observed façade concerning its basic façade elements and their interrelationships. This knowledge is applied in three ways. First, the façade model generated during the data driven reconstruction process can be verified and made more robust against inaccuracies and false reconstructions due to imperfect data (section 3.1). Second, façades which are only partially covered by sensor data are completed (section 3.2). Third, totally unobserved façades are synthesised by a production process (section 3.3).

3.1 Verification

The result of the data driven reconstruction process, which is the basis for knowledge inference, may contain false façade structures and therefore be partly incorrect. Some of these errors can be eliminated during an iterative grammar based verification. For this purpose, a rectified image of the façade is used. The grammar is applied to generate hypotheses about possible positions of each geometry tile and project them onto this orthophoto. An image correlation process decides whether a proposed position is accepted or rejected. In case of acceptance, the geometry can be inserted; existing geometries that intersect with the new one are deleted. Afterwards, the resulting improved façade model is used to update the set of terminals and production rules.

Fig. 3 depicts the orthophoto of the data set "Prinzenbau, Stuttgart" as well as parts of the reconstruction result before and after verification. The grilles of the arched windows in the ground floor cause reconstruction errors (Fig. 3b). Only the window to the right of the door could be reconstructed correctly. Its corresponding image mask and the hypothesised and accepted positions in the orthophoto are marked by a yellow rectangle and white crosses, respectively (Fig. 3a).

3.2 Completion

Due to the scan configuration during data acquisition, façades may contain areas where no or only little sensor data is available. In such regions, an accurate and reliable extraction of windows and doors cannot be guaranteed anymore. Nevertheless, a grammar based façade completion allows for meaningful reconstructions even in those areas. The main idea is to derive the façade grammar solely from façade parts for which dense sensor data and thus accurate window and door reconstructions are available. The detection of such 'dense areas' is based on a heuristic approach which evaluates the sampling distances of the points lying on the façade surface. The

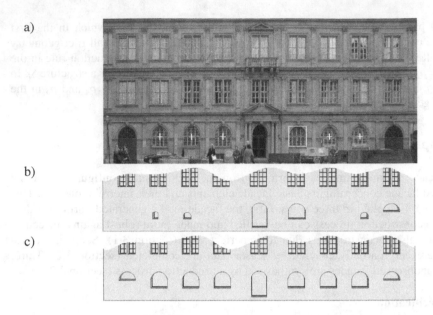

Fig. 3. Prinzenbau, Stuttgart: Orthophoto (a), 3D model before (b) and after verification (c)

restriction of the grammar inference to dense areas ensures a façade grammar of good quality. It can then be used to synthesise the remaining façade regions during a production process.

3.3 Production

The production process starts with an arbitrary façade, called the axiom, and proceeds as follows: (1) Select a non-terminal in the current string, (2) choose a production rule with this non-terminal as predecessor, (3) replace the non-terminal with the rule's successor, (4) terminate the production process if all non-terminals are substituted, otherwise continue with step (1).

During the production, non-terminals are successively rewritten by the application of appropriate production rules. When more than one production rule is possible for the replacement of the current non-terminal, the rule with the highest probability value is chosen. As soon as the façade string contains only terminals, the production is completed and the string can be transferred into a 3D representation.

4 Quality Evaluation

The façade models derived in previous sections represent 3D geo data that feature geometric and semantic properties. They consist of 3D cells where each cell is assigned a special meaning, for example 'façade-cell' or 'window-cell'. Both the geometric and semantic aspects of the façade models are uncertain depending on the quality of the sensor data and the reconstruction approach applied. Estimating the quality of such façade models is important, especially when they are to be combined with other data or used within subsequent applications. In principle, two basic strategies for quality assessment can be distinguished: external and internal evaluation.

External evaluation is based on controlled tests using simulated or real data. Reconstructed objects are compared with reference data of superior accuracy yielding quality estimations that are independent of the reconstruction approach. This is necessary to prove the accuracy potential of the algorithm or the adequateness of the applied model. Existing approaches for quality evaluation of 3D building models use either manually generated 3D models or independent measurements with a low uncertainty as reference data [6], [7], [8].

Internal evaluation, also referred to as self-diagnosis, means the estimation of geometric and semantic quality aspects of objects during the object generation process. This requires additional information, for example redundancies in the underlying data or a priori knowledge about the generated object in terms of geometric, topologic or semantic criteria. However, knowledge inherent in the object model can only be applied for an unbiased evaluation if it has not already been used within the reconstruction [9]. Thus, self-diagnosis is strongly linked to the object generation process. It achieves autonomy within a chain of automatic procedures where generally no reference data is available [10]. Error propagation applied to the stochastic properties of input data results in precision measures which are appropriate indicators of the accuracy if the estimation model can be considered correct.

We aim at the quality assessment of 3D façade structures as derived in previous sections for both bottom-up and top-down reconstruction. Due to the lack of highly detailed reference models, external evaluation based on real 3D buildings is not possible. On the other hand, internal evaluation is only appropriate for reconstruction processes that can be described mathematically and thus are the basis for applying error propagation. Though this is the case for the bottom-up reconstruction (section 2.1), the top-down modelling (section 3) rather represents a black box instead of a clear functional model. Thus, we strive for different approaches for quality estimation depending on the reconstruction algorithms proposed. While this is still ongoing work, first considerations are presented in sections 4.1 and 4.2.

4.1 Quality Evaluation for Data Driven Reconstruction

The data driven façade reconstruction (section 2.1) is based on cell decomposition. The delimiters of the cells are planes which are determined through least squares adjustment from LiDAR points at window borders. Thus, the quality of the 3D façade structures is mainly influenced by the uncertainty of the estimated planes and the applicability of the model assumptions inherent in the algorithm.

The prediction of the uncertainty in the plane parameters requires information about the stochastic properties of the LiDAR points. Two main error sources affect the quality of the laser data: the uncertainty due to the georeferencing procedure, and the point noise inherent in the scanning process. The accuracy potential of the georeferencing of terrestrial laser data depends on the scan configuration, the availability of control and tie points as well as the accuracy potential of the sensors [11]. However, within the area of interest, which is a single façade, each of these error sources has the same influence on the LiDAR points. They act as systematic errors in the least squares adjustment without any stochastic effects on the quality of the plane equation

[12]. When focussing on the relative accuracy of façade structures instead of the absolute one, these factors can be neglected.

The uncertainty of the plane parameters purely caused by point noise can be estimated through the least squares adjustment in terms of the cofactor matrix for the unknown plane parameters: $M^{-1}=(A'Q_{ll}^{-1}A)^{-1}$ where A stores the functional and Q_{ll} the stochastic model of the plane estimation. Geometrically, the matrix M defines an elliptic bipartite hyperboloid which is the uncertainty region of the estimated plane. Its parameters can be used as quality measures for the planes. Error influences based on inappropriate model assumptions have to be considered separately, which will be part of our future work.

4.2 Quality Evaluation for Grammar Supported Reconstruction

The proposed method for top-down reconstruction aims at the modelling of façade parts for which no or only a few point measurements are available. It is based on grammar rules which are automatically derived from façade structures reconstructed during the bottom-up process. The application of the rules is guided by a stochastic process and thus cannot be expressed in functional terms. Consequently, quality evaluation requires external reference data. Since, in our case, highly detailed 3D reference data is not available, we use additional image data showing the reconstructed façades in reality as additional, independent observations. Images are converted into binary images in which wall regions are distinguished from façade structures such as windows and doors. Additionally, similar binary images are generated from each of the modelled 3D façades. Corresponding images are compared by means of image correlation. The resulting correlation values describe the difference between the 3D model and the information derived from real data. Hence, it can also be interpreted as a quality estimate for the reconstructed façade.

This quality measure is just one possibility to roughly assess the accuracy potential of the grammar supported reconstruction process. Other evaluation methods are currently analysed in order to find the most meaningful quality estimation. Furthermore, future work will go into more detail by searching for factors that influence the quality of the façade model. It will be examined whether certain properties of the grammar can be identified as affecting this quality significantly.

5 Results and Conclusions

LiDAR data as it is used within our data driven reconstruction process is usually acquired by either static terrestrial laser scanning or mobile mapping from vehicle based systems. Despite the good geometric accuracy, which can be realised by mobile mapping systems [13], the unavailability of measurement data for building parts is particularly a problem with these systems since only street-facing façades can be observed. Fig. 4a depicts the configuration for the "Lindenmuseum, Stuttgart" where the LiDAR points are obtained by the mobile mapping system "StreetMapper" [13] for a single façade. The direction of driving is marked by a red arrow.

Restricted to the extent of those measurement areas, the output of the data driven bottom-up modelling is used to complement the rest of the building. Fig. 4b and Fig. 4c show the resulting 3D model from different viewpoints. Although the building is quite complex, geometrical structures derived from the street-facing façade (yellow shaded area in Fig. 4c) are sufficient for synthesising the whole building. While in this example at least one façade could be observed completely, typical problems arising with data acquisition by mobile laser scanning are demonstrated in Fig. 4d. Narrow streets with high buildings inevitably lead to oblique viewing angles and thus insufficient point densities in upper building parts. The office building in Fig. 4d represents such an example where only the lower floors could be reliably reconstructed based on LiDAR data. Nevertheless, using the grammar derived from the marked region, the façade and the superstructure on the roof could be completed.

Thus, due to the combination of bottom-up and top-down modelling, our approach is highly flexible towards data of different quality. In this regard, the completion of façade structures at areas of limited sensor quality has been demonstrated. Moreover, façade reconstruction is also possible for whole districts featuring uniform architectural styles if a small set of façade grammars is derived from just a few observed buildings. In addition to the good visual quality, which could be realised for all of our reconstructed 3D models, a number of other applications using the resulting building structures will require a decent quality assessment of the results. While different strategies already were proposed within the paper, this issue is still ongoing work.

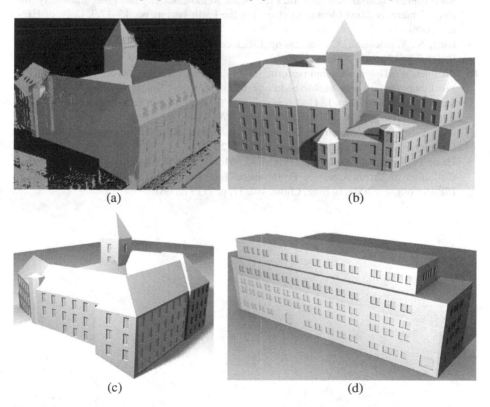

(a)   (b)

(c)   (d)

Fig. 4. Coarse building model and available LiDAR points for the Lindenmuseum (a), results from grammar supported reconstruction for the Lindenmuseum (b,c) and an office building (d)

References

1. Becker, S., Haala, N.: Refinement of Building Façades by Integrated Processing of LIDAR and Image Data. In: IAPRS & SIS, vol. 36(3/W49A), pp. 7–12 (2007)
2. Müller, P., Wonka, P., Haegler, S., Ulmer, A., Van Gool, L.: Procedural Modeling of Buildings. ACM Transactions on Graphics (TOG) 25(3), 614–623 (2006)
3. Bekins, D., Aliaga, D.: Build-by-number: Rearranging the real world to visualize novel architectural spaces. In: IEEE Visualization, pp. 143–150 (2005)
4. Müller, P., Zeng, G., Wonka, P., Van Gool, L.: Image-based Procedural Modeling of Façades. ACM Trans. Graph. 26(3), article 85, 9 (2007)
5. Van Gool, L., Zeng, G., Van den Borre, F., Müller, P.: Towards mass-produced building models. In: IAPRS & SIS, vol. 36 (3/W49A), pp. 209–220 (2007)
6. Henricsson, O., Baltsavias, E.: 3-D building reconstruction with ARUBA: a qualitative and quantitative evaluation. In: Automatic Extraction of Man-Made Objects from Aerial and Space Images (II), Ascona, pp. 65–76 (1997)
7. Akca, D., Freeman, M., Gruen, A., Sargent, I.: Quality Assessment of 3D Building Data by 3D Surface Matching. In: ISPRS Congress Beijing 2008, Proceedings of Commission II, vol. 37, Part B2, Com. II, p. 771 (2008), ISSN 1682-1750
8. Freeman, M., Sargent, I.: Quantifying and visualising the uncertainty in 3D building model walls using terrestrial lidar data. In: Proc. of the Remote Sensing and Photogrammetry Society Conference 2008 Measuring change in the Earth system, pp. 15–17.9. Univ. of Exeter (2008)
9. Hinz, S., Wiedemann, C.: Increasing Efficiency of Road Extraction by Self-diagnosis. PE&RS 70(12), 1457 (2004)
10. Förstner, W.: Diagnostics and performance evaluation in computer vision. In: NSF/ARPA Workshop: Performance vs. Methodology in Comp. Vision, Seattle, WA, pp.11–25 (1994)
11. Schuhmacher, S., Böhm, J.: Georeferencing of Terrestrial Laser scanner Data for Applications in Architectural Modeling IAPRS, Part 5/W17, vol. 36 (2005)
12. Oude Elberink, S., Vosselman, G.: Quality analysis of 3D road reconstruction. In: Laserscanning 2007. IAPRS & SIS, Espoo, Finland (2007)
13. Haala, N., Peter, M., Kremer, J., Hunter, G.: Mobile LiDAR Mapping for 3D Point Cloud Collection in Urban Areas - a Performance Test. In: IAPRS & SIS, Part B5, Com. 5. ISPRS Congress 2008, Beijing, China, vol. 37, p. 1119 (2008)

Author Index